On Spirits

The World Hidden Volume II

Dr. Joseph Michael Feagan

For my wife and children.
If the darkness is real, so too must be the light.

Table of Contents

Preface

This book is all about *so-called* spirits, duh. The title gives it away, and purposely so. I needed there to be no ambiguity about it, not even a little bit. But instead of bombarding the reader with stories of the sensational, those skin-tingling, creepy-crawly, chills-engendering, keep-you-up-at-night kinda stories, this is a book primarily about how the existence of unseen intelligent entities can be proven scientifically. The journey of experiences which helped me achieve that gargantuan scientific milestone is recorded here, at least in part, and it was certainly not a feat achieved by my own efforts alone. Make no mistake about it, there are definitely some things, some weird but true accounts that will challenge your credulity in the pages ahead. And some of them will grab hold of you as if by the neck and won't soon be forgotten. I had to put some of that in the book to keep the pages turning, plus I thought it important to show how I painstakingly found my way forward scientifically largely through research and interviews with individuals with intriguing personal experiences with spiritism.

Science, by its very nature, isn't everyone's cup of tea, and unlike many scientific breakthroughs, some of which are too sophisticated for the untrained to understand succinctly, the existence of spirits is a concept that even a child can understand. Their existence, too, has many practical real-world applications which should be openly discussed and considered, as well as vitally important religious considerations. And in this book, we will delve into those philosophical and religious considerations as well. This is essential for many reasons, not least of which is that ancient cultures often possessed profoundly advanced cosmological viewpoints that were expressed through the language of mysticism. For example, something that resembles what science now calls string theory has been expressed in various Indigenous cultures for millennia, long before it appeared in the modern Western scientific lexicon.

For individuals from lands familiar with ancestor veneration, those experienced with what is routinely referred to as the occult, and others who perhaps were raised around various types of spiritistic practices, this book's authenticity will immediately resonate with you. For others, be prepared, much as I was, to have a whole new world, a completely hidden world for many Westerners, openly exposed for you to ponder and consider.

To be clear, because of my own personal beliefs, I do not personally advocate for participation in spiritism or occult practices in any way, nor do I personally engage in any such conduct myself. Within the confines of the pages that follow, I write as a journalist, physician, and scientific theorist and have intentionally withheld my own personal feelings about the morality and ethics of such practices, other than to have a philosophical discussion. Life has repeatedly reminded me, especially as a physician, and many times as a journalist, it's not my place to judge the practices and behavior of others. To be a rational and dispassionate observer, simply thinking about and stating the ramifications and implications of spiritism is my chief aim within this work. I also realize that there are physical-world ramifications intimately linked with various practices associated with spiritism, and if the mechanism of such physical-world alteration is fully understood, it will have enormous practical scientific ramifications. For this reason primarily, I do advocate for the scientific study of these practices, which no manner of religious-based objection, or prejudice-based persecution of individuals who engage in various spiritistic practices, will prevent from being completed in the world as it now exists.

The existence of unseen intelligent entities can easily be proven with very simple modern scientific methods. It behooves humanity to understand and acknowledge the existence of such beings, for comprehending how energy can be organized in such a way as to enable consciousness and intelligence is a path that may potentially lead to previously unconsidered forms of renewable clean energy and many other important scientific endeavors. The study of unseen entities is a vital, relatively unexplored path, and aside from any spiritual and religious ramifications such study engenders, there is absolutely no way to predict how significantly humanity's technological progress will be altered through the study of this very real form of intelligent alien life. So sit back, relax, and prepare to have your mind blown—with all due respect, unseen intelligent beings do exist.

Introduction

While it may not come as a shock to you, some of history's greatest figures owe an enormous debt to their own subconscious minds. I can say with complete clarity of recall, I certainly never knew anything about the importance dreams have played in scientific discovery repeatedly throughout history. As it relates to scientific discovery, whether people realize it or not, some of the world's greatest scientific and artistic achievements were influenced partly and even entirely by that subconscious wonderland, the dream state. Imagine, for a moment, having the fortune of becoming rich and famous in no small measure because of the insight offered by a single vivid dream. It's difficult to imagine, but it is in fact true, and it's happened several times throughout humankind's history. Some even go so far as to say it's as if some unseen force was maneuvering the development of humankind's technological advancement through the subconscious minds of humanity's greatest thinkers. Your viewpoint of that statement may entirely define the lens through which you view the information that follows herein.

Albert Einstein knew the importance of dreams and intuition in a poignant way perhaps only a few people do. His own theory of relativity was born from an epiphany a dream about cows and a farmer birthed within him. That's right—a dream about cows birthed one of the greatest discoveries in human history. Niels Bohr, a contemporary and associate of Einstein, and a Nobel Prize–winning chemist himself, also had his own Nobel Prize–winning theory inspired similarly by a dream. In his dream, Bohr saw the planets of our solar system attached to pieces of string circling the sun. From this, he envisioned the movement of electrons, and like Einstein, his theory would impact humanity's understanding of the universe profoundly. This list goes on and on, and while it may not surprise you personally, many of history's great discoveries and figures, including the rationalist Descartes, Mendeleev's periodic table, Mary Shelley's *Frankenstein*, Paul McCartney's "Yesterday," and other profound inventions and artistic creations, all owe their inspiration, at least in part, to that subconscious wonderland, the dream state.

Despite the bespectacled stereotype of scientists wasting away in labs, many of the world's most significant scientific discoveries, songs, stories, and inventions didn't happen in a lab, or in front of a computer screen or microscope, at least not in the way people often perceive great discoveries are made. And the discovery of alien life, too, didn't happen in the way it is often depicted in cinema. Yes, my own journey of scientific breakthrough was anything but traditional, and the path to verifying the existence of alien life was an equally atypical and circuitous one for me.

Initially, born by my own unique experience with a shaman, a unique dream, and then by journalistic leads resulting in gathering multiple eyewitness-confirmed reports of supernatural occurrences, I realized unequivocally that unseen entities, "spirits," are in fact a very real actuality. But how to go about proving their existence scientifically? That was the central question for me. It wasn't enough to just believe and personally know they're real. Many people feel that way. The real step, the gargantuan significant one for humanity as a species, rested in establishing a repeatable, quantitatively measurable means of proving their existence. This was my task, one that felt more thrust upon me by what some might call destiny, or the divine force, than one personally chosen. And my own curiosity insatiably fueled a search for a path forward to prove that unseen entities are in fact verifiably real.

As a student of history, and of science and religion in particular, I realized the importance of looking beyond the surface story, for the truth is often hidden in those smallest of seemingly insignificant details. Interestingly, it was in the gathering of these eyewitness accounts, the persnickety perseverance in building relationships, talking, and visiting with individuals often considered unintelligent and poorly informed, that the real discovery of alien life was made, and my path forward elucidated.

At its heart, this is what *On Spirits* is all about—the captivating and circuitous path to discovering alien life—and it was nothing like what the movies and traditional thoughts about such life have customarily centered around. Principally, in writing this book, I wanted people to share the value of the insight and wisdom those who shared their personal experiences with me imparted. Their candor, their authenticity, their humility and kindness in sharing deeply personal accounts with me moved me in such a way that I was convinced of the accuracy and veracity of their experiences, so much so that an insatiable drive lit within me. Combined with a clarity of purpose, I knew intuitively there must be a way to verify scientifically what they had subjectively experienced. Perhaps the single greatest reward for me personally, aside from the sheer awareness of what the staggering significance of proving the existence of alien life means for humanity, is that my discovery in large part was aided by the help of others without the assistance of an enormous budget or cutting-edge technology. The verification of alien life didn't require satellites, radio telescopes, or the hundreds of millions of dollars spent each year scouring the galaxy; rather, the simple study of water, fueled by conversations with individuals often flippantly dismissed as ignorant, was all that was required.

I've come to know many a sleepless night since my discovery, and my wife and children have suffered as much as I have as my unique journey increasingly consumed more and more of my time.

Five books in the last three years and my storage shed morphed into a makeshift laboratory also bear testament to my labor. But there are simply some callings that beckon sacrifice. And whether it be time or money, sleep, or something else, there are moments when the pursuit becomes the all, and the call to action simply cannot be ignored. For me, that moment first came in the summer of 2019, on my third trip to Cuba, when for the first time I spoke and interacted at length with a genuine shaman. While my experience was not unique, in that countless people from various cultures have experienced something similar, the epiphany it birthed within me posed consequential ramifications for the whole of human society that cannot be quantified. The shaman told me things about myself that even my wife and family didn't know and knew things about my family that even I didn't know at the time, but I have since confirmed as being accurate. I am both a physician and a journalist, and logic and reason bade me to understand scientifically the nature of what I'd experienced. For me, a rationalist, there was simply no such thing as magic, in the sense that no phenomenon happens without a scientific explanation. Paranormal and psychic experiences are uniform to every culture that has ever existed in the history of humanity. But there are things, despite our many scientific advancements, that can neither simply nor complexly be explained with our current degree of scientific understanding. Should we simply dismiss actualities we cannot understand, things for which there is no current rational explanation, or concepts that contradict our preconceived, often deeply ingrained biases?

The nature of scientific exploration is to voraciously pursue an understanding of phenomena that defy our current paradigm of scientific knowledge. To embrace those boundaries, explore and breach them, has always been the fuel that has expanded the consciousness of human technological advancement and human civilization. In this regard, the study of unseen entities and those who interact with them is as rational an area of scientific pursuit as any other.

After my initial interaction with that shaman, I realized her ability wasn't rooted in intrinsic human ability alone, as she herself acknowledged her insight was the product of an ability to interact with unseen entities, who have heretofore been commonly referred to as spirits. Throughout history, in every culture that has ever existed, shamanic-type figures have persisted within various Indigenous and other communities who claim to interact with unseen entities. Cultures separated by enormous geographic distances, and even centuries of time, who share common rituals, practices, and beliefs engender sincere questions as to why such congruence of accounts persists. There are only two explanations: either (1) unseen entities, commonly referred to as spirits, are real and the nearly parallel accounts of them, and shamans' manifestation of unique and diversified abilities uncommon among humanity, are accurate, or (2) nearly every culture that has ever existed in the history of human civilization is plagued by the exact same delusion.

For centuries, modern scientific theory has nearly exclusively viewed such accounts, and the cultures that contributed them, through the lens of the latter perspective. Compelled by the indoctrination of Abrahamic religious traditions, media bias, racial prejudice, misunderstanding of Indigenous peoples, and simple ignorance regarding the reality of these entities' existence, science has failed to accept, study, and understand what countless Indigenous people have understood for centuries. Driven by an unequivocal understanding of the existence of these entities, I began interviewing first family members and friends in Cuba. Cuba, like many other impoverished nations, has a rich history of individuals both practicing as and interacting with shamans. Ironically, my wife is a descendant of several generations of Cuban shamans, and later, other acquaintances and journalistic leads throughout the United States contributed their accounts. I interviewed them, obtained multiple eyewitness testimonies when possible, and attempted to confirm as many of these difficult-to-believe accounts as my sources provided. I am not a man inclined to credulity, and given the spectacular nature of many of the accounts I've confirmed by multiple eyewitness sources, I realized that simply acquiring and confirming such accounts would not be enough to convince the general Western public of these accounts' accuracy. Had I not experienced such things myself directly, even I wouldn't be inclined to believe them.

By training and experience, I am a competent photojournalist, with a twenty-five-year history of traveling throughout the United States and the Caribbean documenting primarily the life of the "ordinary person" and the impoverished. Two and a half decades of connecting with people through the lens of my camera, documenting some of their most sensitive and vulnerable moments, truly private aspects of their lives, and sharing their stories in a thoughtful and considerate way have been the fuel keeping balance in my other life as a physician, husband, and father. The individuals I spoke with following my transformative shamanic experience afforded me unique and unfettered access to intensively private aspects of their lives, particularly rituals and practices that customarily have been hidden intentionally. Because of the sacredness with which their activities are perceived and the sensitivity of what is involved—the use of human bones, for example, in some traditions—it's obvious why the details of what many practitioners do are kept secret. Two and a half decades of telling people's stories have taught me there are gifts within the field of journalism that simply cannot be taught, for true journalists are born, not made in my opinion.

The hunger to learn something new, to sincerely listen to the stories and experiences of others, especially the vulnerable and often ignored, and to tell those stories with honesty, integrity, respect, and compassion for the individuals opening themselves up to you, is a sacred trust. This is the most beautiful gift of journalism. To get to know a small piece of a person, to cherish that piece and share it with the world so that others too may share in savoring the beauty and struggle of those with experiences that impel one to tell, is a joy and privilege of inestimable value.

The World Hidden, my first book, was published in April 2021 and featured twenty-five firsthand accounts, the majority with multiple eyewitness confirmations of the various activities of modern shamans, primarily in Cuba and the United States. It was a tremendous disappointment to me that that book didn't include shamans from around the globe, as shamanic cultures persist throughout the world. In the past two years, I've attenuated my practice of medicine to write and have written several books: the book previously mentioned, *Love, The Illusion of Superiority*, and now the book you're currently reading.

The nature of scientific discovery is such that observation is customarily the first step to achieving progress of any significance. In my travels in pursuit of the truth regarding unseen entities, and in the activities of various shamanic practices and beliefs, I observed that a glass of water is commonly placed on the altars of practitioners of various traditions. Over time, this water develops a unique opacity, and a transparent, perfectly appearing sphere forms just below the surface of the water. I became convinced that the physical analysis of this shamanic water would provide unquestionable, repeatable, quantifiable measures that conclusively establish the existence of these entities and have already completed experiments that show early unequivocal results within this field of inquiry. The combination of this analysis of shamanic water with modern-day biometric analysis, including brainwave analysis, various types of thermal and spectral imaging, and radiographic imaging of patients treated before and after shamanic intervention, forms a comprehensive way of assessing the abilities of shamans and both indirect and direct proof of the existence of unseen entities.

Without such proof, no number of confirmed eyewitness accounts or video footage will be able to convince conclusively the astute and fiercely scrutinizing scientific community, which will no doubt rigorously refute the claims and testimony made within the fabric of my written work concerning the reality of these entities. I welcome such, with full confidence that with persistence and support, the evidence of what Indigenous people have known for thousands of years, and developed rituals and systems of practice to use for the benefit of their communities, will be established. Unseen entities—spirits—are real, and their existence can be proven scientifically. Obviously, the nature of identifying alien life, and the enormous religious and technological implications of such, inevitably places the value of my work in the highest echelons of significance.

While basic scientific research is ultimately how the existence of unseen entities will be conclusively established, the pathway of which I outline in detail in a later part of this book, the acquisition of multiple eyewitness accounts from both shamans and others who both interact with and have observed the activity of these entities formed the foundation of how I understood that pathway. Conclusively proving the existence of the entities rests in the simple yet comprehensive analysis of water.

In an effort to outline my journey to proving the existence of unseen entities, *spirits*, this work takes the reader through my own experiences, including reference materials and research materials that aided me in my discovery. As many a great discovery often rests upon simply building off the work of previous scholars and others, so too my own was one not born solely of my originality, for a great number of both healers and others helped to guide my thoughts and observations. My disdain for both the typical language and tenor of scientific writing obliged me to formulate something that can be easily read and understood by a large audience of readers, hence this book's unique combination of eyewitness experiences and essays all centered around my effort to scientifically explain the precise nature and activity of these entities.

While the subject matter of this work will certainly challenge readers' credulity, my observations and my work with shamanic water are accurate, and in the near future, I assuredly believe such work will be easily verified by multiple scientists, and the challenging work of where we go from here can begin, as the ignorant denial of the existence of spirits will finally and irrevocably be removed from the consciousness of large segments of our civilization.

The Rediscovery of Alien Life

*"The truth is like a lion, you don't have to defend it, you just
have to set it free and it will defend itself."*
—Anonymous

*"Water symbolizes the whole of potentiality—the source of
all possible existence."*
—Mircea Eliade

Everybody always wants to be first. From the first one picked for kickball in elementary, to the first one with a car in high school, and the first one on and off a plane—and everything else in between and after. Passivity and patience are certainly frowned upon as weaknesses in this world where so many believe the quality of their life rests on being the first and best. Being first certainly does have its perks, though. From being the most popular, to getting laid, and even becoming rich, depending on the thing you're first in, the ramifications can literally be life-altering and, at times, even history-changing. Still, despite the pressure-driven realities of doing whatever it takes to be first, some parts of making history simply cannot be rushed.

Dr. Joseph Michael Feagan

The notion a person can just make some outlandish statement and have it accepted as fact has become an actuality no longer rooted in incredulity. Social media, in particular, alongside mainstream media outlets far removed from the days of neutral-leaning content offered without bias, has made what were once heretofore statements of fake news not only commonplace but the new norm in an age where the quest for fame, wealth, and political power has placed the values of integrity and honesty in inferior positions. Even within the realm of science and healthcare, often the results of scientific experimentation are vigorously refuted or denied largely based on artifice, simply a displeasure at what such studies reveal to be factual. The recurring episodes of climate change deniers easily illustrate such a pattern in the modern era. Still, what is, is, and perhaps in no greater field of endeavor has that actuality remained more firmly in place than in the realm of our scientific understanding of the universe.

Gravity just doesn't go away, no matter how much denying or gaslighting you do. Go ahead and try to deny gravity, say by stepping off a ledge, and you'll feel just how quickly science slaps you in the face with a hefty bit of reality check. Photosynthesis, principles of electricity and magnetism, and other bedrock foundations of our modern understanding of the universe simply cannot be trivialized no matter what efforts are made to refute them. That, above all else, is the true force of scientific fact. What is, is, and science remains the last absolute pillar of truth remaining in society during an era when integrity and journalistic ethics are increasingly tossed aside in favor of the bombastic and controversial. What is, no matter how voraciously it is assaulted, cannot be stopped, bent, or broken. And it is this great comforting actuality that steadies my own firm conviction that what I have achieved will in time be accepted as that which it is, the irrevocable and undeniable proof that unseen entities, *spirits*, both exist and have interacted with humanity for millennia.

For a scientifically minded individual to be confronted with the possibility that unseen entities exist is a challenging notion. Aside from the staggering scientific significance, the existence of unseen entities forces one to entertain concepts and ideas that can be extraordinarily challenging to consider for a closed mind. For such a mind bent on holding cherished perceptions of the universe and all that such comprises, the existence of unseen entities engenders a host of questions inextricably linked to religious concepts often flippantly denied by the most accomplished of scientific minds. Still, I ask the reader to entertain and accept for just a brief moment that the lexicon much of the ancient world possessed had no word for alien or extraterrestrial. In this vein, the language of the mystical and religious became the language for explaining all things in the realm of the scientifically unexplainable in many previous eras. Reflecting on the limitations in lexicon and language of many intelligent but less scientifically advanced civilizations and peoples makes all the difference when viewing these ancient people's accounts regarding phenomena they experienced.

It is a grave mistake to think that humans today are somehow more in tune with the natural world or invisible realities humanity poorly understands. In fact, quite the opposite is true. Many ancient people understood far more about the unseen realities I will be discussing within the body of this work, and their opinion on such matters should not be ignorantly disregarded. Imagine for a moment attempting to describe an airplane to someone who lived two thousand years ago. Such would be no easy task, in large part because of the ancient world's limitations of vernacular and scientific understanding. While a rudimentary word, *spirit*, engenders a nearly universally embraced concept by cultures around the world. It is an ancient word that still perhaps best describes how these cultures experienced phenomena not easily explained. They had an awareness of these entities' existence, their intelligence and consciousness, and yet little to no scientific understanding of what they were. Still, with such limitations, many cultures around the world developed systems of practice for interacting with these entities with remarkable similarities.

As is to be expected, others—outsiders routinely excluded from traditions and practices frequently kept hidden—have noticed that these systems and certain behaviors often associated with them display characteristics and happenings that engender sincere questions from rational observant minds. Many a great ethnographer and scholar have dedicated their lives to understanding Indigenous peoples and their patterns of practice in interacting with these entities. To such scholars and writers as Lydia Cabrera, Mircea Eliade, and many others, I and human civilization as a whole owe a debt that can never be repaid. In a sobering yet poignant cognizance, I understand sincerely and deeply, without their own trailblazing scholarship and devotion, my own discovery would never have been made.

In the pages that follow, I take the reader on both a scientific and a spiritual journey unlike any other they've likely known before, through the lens of the experiences of other sincere observers, traditional Indigenous healers, scholars, and others. For, despite the often-depicted antagonistic relationship between science and religions, they are, in fact, intricately interwoven. This actuality exists primarily because the way humankind has typically dealt with scientific realities before understanding them has, nearly without exception, been through the lens of religion and myth. Tornados, lightning, hurricanes, tsunamis, and other natural phenomena, before science did the detailed work of explaining them, all had their depictions relayed in mysticism, story, and art. And until more modern times, the world perceived such phenomena as actualities, long before they were explained scientifically.

Herein lies the primary way in which religion, mysticism, and however else you want to label humanity's pursuit of the ultimate higher life form unseen should be considered. Humanity's quest for the ultimate higher life form has always been the primary language through which legitimate observations of the unexplainable have been described before their scientific explanation. And, similarly in this regard, the nature of unseen entities is in fact no different. Instead of the typical approach science has made in disregarding religious and Indigenous people and their cosmology and perception of the universe and *spirits*, I took a step back, fueled by the rationality of my own difficult-to-explain personal experience, and considered for a moment they are accurate in their depictions of intelligent unseen beings. And then I considered: how would one possibly go about proving their existence?

Just like lightning, hurricanes, tsunamis, and other difficult-to-explain natural phenomena were first perceived through the lens of mysticism, so too, perhaps, patterns of practices, customs, and rituals are the vernacular through which legitimate, quantifiable experiences are taking place. I am far from the first to understand, perceive, and pursue a clear and precise understanding of the nature of unseen entities. Countless Indigenous peoples and many other individuals throughout humanity's history, including my own ancestors, have both known and understood the reality and the existence of unseen entities and additionally developed patterns of practices to interact with these entities. My sober acknowledgment of this actuality is the reason for labeling this chapter the *rediscovery* and not the *discovery* of alien life. For while my own awareness of the significance of water in identifying these entities represents a milestone for humanity scientifically, my ancestors and others have long understood the existence of unseen entities and their association with humanity, and their contributions in this regard I cannot and will not dismiss or ignore.

So sit back, and try to keep an open mind. Know that the words that follow, no matter how difficult to believe, are indelibly rooted in a truth so staggeringly significant, it is posed to change the course of human civilization. Nobody reads scientific books for fun—come on, let's just be honest about it. And I got mouths to feed, dude, four of them. So firstly, not that I could even do it, but I really couldn't stomach making a book so brilliant only an Oxford or MIT grad student or PhD could understand it. What would be the point of that? The essential idea, and the obvious one at that, is not simply presenting something fascinating to enhance humanity's understanding of the universe, but the practical ways such knowledge affects people's lives in the *real world*. In this vein, the journalistic efforts that aided the discovery of alien life are important for people to read and experience for themselves.

One of the aspects of my work that brings me great satisfaction, particularly as I've had time to process the significance of what identifying alien life means for humanity, is the beautiful and inspiring stories I've listened to and gathered. These accounts both educated and inspired me, and shaped my practical understanding of what the existence of unseen entities means, and what it has meant in a practical way for millennia. Countless Indigenous and other peoples have indeed used their understanding of these entities to assist their communities for generations. In putting this work together, I wanted readers to experience the journey my own life took as a consequence of these unique experiences and accounts, and how these culminated in me discovering a path to scientifically verifying the existence of unseen entities.

I've always been both a student and a fan of scientific giants like Bohr, Einstein, Pauling, Carver, and others but frequently found myself wishing I knew more about their journeys to their discoveries. Their thoughts along the way, their failures, their persistence often get pushed aside and the achievement set apart without any real understanding of the determination and grit that contributed to how it was made. There is often this air of impossibility surrounding great discoveries that frequently adds to the mystique of many of science's great minds, but what I've observed in my own life, at least, is that while brilliance is certainly a helpful asset in life, a hefty dose of good fortune and a persnickety persevering curiosity seem the far more important ingredients in the achievement of anything truly trailblazing scientifically.

For me, my journey to identifying alien life took both a very linear route and a serpentine and circuitous one. Initially, I had an experience I couldn't explain logically, so I pursued understanding it linearly. I sought others who've also experienced or observed similar things, confirmed with multiple eyewitness accounts the proliferation of such phenomena, and then, over time and with observation, formulated and implemented scientific experiments to test hypotheses rooted in the observation of patterns of similarity found in my journalistic and scholarly endeavors. While the discovery of alien life certainly belongs among the highest echelons of achievement, it was a path on which I received an extraordinary degree of support and assistance. In fact, and with great certainty, I know I wouldn't have made any significant progress in this endeavor were it not for the loving support of many gifted healers, friends, and family, as their own experiences, observations, and accounts, in a prodigious and elucidating way, gradually helped me to identify a path forward of analyzing shamanic water in a scientifically repeatable and quantifiable way. This in no small part was aided by my own dedicated path of routine religious study and curiosity, not merely as a spiritual exercise, but as a study in cosmology, where religious concepts of cosmology are frequently dismissed flippantly by profound scientific minds without considering that there are perhaps aspects of religious study that hold enormous promise for greater understanding of the universe scientifically.

How is this book organized? Hopefully cool, inspiring, and insightful stories—the same ones that flipped my wig initially and led me to keep digging for a path forward scientifically as it were. Interspersed with those, what I think are just as cool and insightful text: my own thoughts and experiments on unseen entities. Toward the latter part of the book, I detail the experiments that form the foundation of my discovery and the future experiments expected to further elucidate the heretofore nebulous world of unseen entities. Whatever follows, perhaps even if shrouded in incredulity, I pray humanity finds the courage to unmask the bigotry and ignorance that for far too long shrouded the world in disbelief of a reality so significant, it will change the world forever once globally accepted—**we are not alone in the universe!**

Accounts

The accounts that follow were obtained through the candid cooperation of various acquaintances, friends, family, shamans, curanderos, santeros, espiritistas, paleros, and others of good repute. The individuals come from all walks of life and have a range of various religious beliefs (or a lack thereof). Atheists, Muslims, Buddhists, Protestants, Catholics, Jehovah's Witnesses, Santería devotees and priests, agnostics, and others served as sources of the material reported in the pages that follow. No payment or remuneration of any kind was offered or promised in the present or future to those who shared their accounts.

These accounts were gathered from both English and Spanish native speakers, and the interviews were conducted in the native languages of the persons whose accounts were taken. In some instances, the interviews were recorded in video or audio, when possible and permitted. In others, the accounts were transcribed from notes. In some cases, multiple eyewitnesses have confirmed an account's authenticity. The symbol *AC** is used next to an account provider's name when a portion or all of the information given in the experience was attested by at least two individuals other than the person telling their experience. In other accounts, only the person sharing the account reported the events as accurate.

At the beginning of nearly each account, I've added a short blurb about the interviewee and the circumstances that led up to the interview. While certainly not essential reading, it's my hope these short additions will add some color and detail for the reader to otherwise faceless contributors.

A Humble Priest, Thousands of Exorcisms, and Two Near-Death Experiences
as told by Dr. Virginus

When I met Dr. Virginus one summer morning in the suburbs of a populous Maryland city, I knew I'd have to mind my Ps and Qs. A close friend and acquaintance of my sister, I knew, quite literally, there'd be hell to pay from her had things been anything less than ideal for her friend as we sat for our hour-and-a-half-long interview session. The slim, dark-skinned Dr Virginus had a calm and dignified demeanor and looked considerably younger than his age. The disfigured nature of Dr Virginus' right eye caught my attention immediately, but perhaps less so than his obvious ease and comfort with himself. I forgot exactly how my sister introduced the subject of the two of us meeting; I mostly just remember her saying, "You have to talk to him." She was familiar with my work, and from her tone and confidence, I knew my time with him would be well spent. An accomplished scientist in her own right, and a magnanimous spiritually minded woman, I knew for her to take the time to recommend him and work to bring us together, his experiences, particularly his near-death experiences and his unusually long tenure as an exorcism-performing priest, would be invaluably insightful. And they certainly proved to be.

Every person has their own very unique experience of God. Even though we may be at church together and hear the same sermon, we don't experience him the same way. Even individuals within the same religion, participating in the same sacraments, may have completely different experiences of God. So, it becomes essential in my ministry to understand, what is this individual's experience with God? God and me, looking at each other—what is my experience of God? And how has that impacted my history, my journey of life, my understanding of reality, my relationship with others? That experience becomes very critical. I'll give you an example. I've always seen Jesus more as a savior because, in the most critical moments of my life, he's been present to save. So while I love Jesus as Lord, I love Jesus as my brother, I love Jesus as a teacher, but what speaks to my person is Jesus as the savior. When I was shot, "How could anyone have been shot in the head at close range and live?" people often ask me—I've been telling that story for years.

From the time I was a small child, maybe four or five, as long I can remember really, I knew the priesthood was my calling. I'd felt, above all, a yearning to relieve the suffering of others, and a fondness for the cleanliness and love only a commitment to serving God gives. But it was a drive and passion that invariably, wasn't one born by my own effort alone.

I was just a small child the first time I saw the glow of godly devotion. A group of seminary students had come to my parish, and one of them had such a glow of cleanliness, that kind that shines from the inside out. It convinced me my destiny was to be a shepherd for God.

I grew up in a household committed to Christian worship in a small village in Nigeria. My parents were nominal Catholics; they were devoted to the church and to God, and our lives revolved around worship. There's a difference between a Christian who just goes through the motions and one who lives their faith. My parents lived their faith. And the example they set every day in that small village as a child showed me the cost and reward of genuine Godly devotion.

My parents had a Godly routine, and it became the guide for the whole family, For me, there was just a triangle: worship, church, and home. That's just who we were, what our life as a family was. But not everyone in our village in Nigeria shared that devotion. Growing up, there was a unique festival in the village. On that day, people from the cities would come to the village, and it was a fun affair with music and dancing, eating and drinking. But my parents were pushing to have it Christianized, because it centered around making a sacrifice to a pagan deity.

At the time, only one person in the village was a professed pagan or traditional worshiper. Everybody else went to church. The belief around this festival was that a deity had been contracted through our ancestors. And every year, the village had to gather to honor this deity to be able to preserve life and bring prosperity to the village. There was this muted conviction that if this deity was not honored yearly, there could be wrath and anger, and people would die, and misfortunes would come to the village, and so on. Now, because so many people went to church, nobody wanted to come out openly and condemn the ritual and stand up for Christ.

So this man, the only professed traditional worshiper in the village, was the one who made the yearly sacrifices to the deity. He would make the sacrifices, kill the fowls, kill the goat, sprinkle blood all over a tree, the day before. And then he would go to this big tree that they always gathered around, because they believed that that was where the deity had been honored initially. He would go there and sprinkle the blood of goats and such. Then, the next day, all the people would come out, and it was all celebration—dancing, eating and drinking, games and fun. The people would say, "Well, we are not part of the sacrifice, so it's okay." But my parents stood against it, and they suffered for it. They indeed were punished for their faith.

While people were gathering, styling in their new clothes, my parents would take us to the church, and we'd spend the entire day in the chapel. We'd carry food, bring our breakfast and lunch, so we weren't tempted to eat anything sacrificed. Their willingness to suffer for their conviction is something I am eternally grateful for.

But we were ostracized for our faith, treated like strangers in our own town. But down the road, a few years later, my parents' faith put an end to that festival because more people started saying, "We can't be serving God and serving something else. We want to follow your family's example." More and more people began to estrange themselves from it, and that was the end of it. But they did suffer social consequences initially for that conviction.

Sometimes the Lord does call to service, and it can be different for everyone, but for me, I could just feel the priesthood was my path. I knew it. At four, I had an experience that still feels just as real today as it did all those years ago. We used to have block rosary centers where children were gathered to pray the rosary. Someone had to go ring the bell to signal it was prayer time, and there was this fire within me—I always loved to go and ring the bell. Once, I asked a grownup, "What is the time?" Of course, I didn't know the time, so I had to ask, and then I would run there to ring the bell to summon others. That day, I went and I came into a small structure, a small building, but it had an altar. So we just gathered there as children. We had a matron, who's almost always a woman or a man, who organized us. And then we'd pray the rosary to the Virgin Mary. We were given extramural instructions, and we sang and were dismissed. This would happen all the time, except one day it was different. I ran as usual to ring the bell, but when I came into that room, there was a woman kneeling down at the altar. She was not a human being. There was light all around her. She had a blue veil. It was the most beautiful sight I've ever seen in my life, and she had the rosary. When I ran in, she turned, and looked at me. The radiance of her glory had me transfixed. I couldn't run out. I didn't know what to do. She didn't say a word. She just looked at me. I didn't know if I should remain there, or if I should run away. It was not scary; it was not fearful. It was just an experience that was completely captivating. . . . I only was four. And as I tell you this, I see it like it was just yesterday.

She never said a word to me, and then like smoke, she just vanished. I fled. I didn't ring the bell. I ran home to tell my mom what had happened to me. And she said, "Oh, you should have knelt down; she would have spoken to you. You saw the Virgin Mary." But it was not until I was seventeen, when I was in high school, that she appeared to me a second time. This time she spoke to me, and she said, "You will work for my son." As soon as she said that, I remembered John, Chapter 2. It said, "Whatever he tells you to do, do it." And at that time, she came to me the same way, with the rosary, and she said, "You will work for my son." I was praying in the house. She appeared to me, and she said, "You will work for my son." And then from that moment, it was like there was a fire within me that became unquenchable. It didn't matter if anybody was going to support me or not—nobody was going to stop me.

Author: *Can you tell me some of the rewards that you feel come from the priesthood? A lot of times, for the average person, it can be difficult to see it as a life that they would want. It's usually perceived from a perspective as something that most people would avoid. What have been some of the most rewarding aspects for you? And how do you feel now about your decision after so many years of service?*

If I was given a zillion opportunities, I think I'd still make the same decision. It's a life of sacrifice, of course. But there is something about being the bridge, connecting people to God, and connecting God to people. One of the rewards I've seen is the privilege of relationships—a relationship with God, relationships with others. And because of my relationship with God, I have something of a sacred trust. People respect this sacred trust, and this gives me access to their most vulnerable moments, to their pain, to their struggles, to their joys. So basically, I am a sojourner on a journey, but I'm also . . . . What do you call people who help others when they go to swim? A life, what?

A lifeguard.

Yes, I'm also a lifeguard. So for me, the greatest reward of priesthood is the relationship with God that now forms you and then informs your relationships with others. You stand at a place of privilege where you can preach the word of God and have people sit and listen, not because you're the most intelligent, not because you're the most handsome of people, not because there is anything of you, but because of what God has made of you. You stand at the altar and you consecrate the Eucharist. You speak to bread and wine, to become the body and blood of Christ. You birth people into the kingdom, through the waters of baptism. This, for me, is not about anything else on the outside. Other simple things come too: the small stipends you get, and other perks. It's not all suffering. People are also appreciative of what you do. You stand in the middle. So you get the persecution, but you also get the joy of serving. For me, that's the greatest reward.

Do you feel like you feel God's support of you on a routine basis?

It's day by day, because I have to continue to fuel that relationship in order to serve faithfully, in order to be the best of who God wants me to be. I'm the chaplain on campus, right now, and I see a lot of people die. Especially during the pandemic, it's been difficult; it was really something terrible. But once, I had an experience. I was called in to come and anoint someone before their passing. I had all this PPE on that suffocates you. I came in, the nurse walked me in, and then she left. When I called the patient's name, she looked up at me, and I spoke to her. I said, "I've come to anoint you. God loves you. We love you," and I anointed her. When I was done with the anointing, I realized that she had passed. That was the first time I was with someone when they passed. I'd never had that experience before.

I held her hand, and it was ice-cold. But what spoke to me was, all the divisions that we create in this world, the political divisions, religious divisions, hey, it's racism, all the things that we battle with. When the chips are down, no one remembers whether you are Black or you're White, whether you are straight or you're bent. It's no longer important. For this woman, I was her only family at the most critical moment in her life. She was White, I was Black. She didn't mind that I came from Nigeria. All that she needed was to get the reassurance that on this transit, God was going with her. I brought that presence. Nothing else mattered.

As I drove home, I began to think about it. I said, "I wish we understood this, that today I became her family." I don't know whether she had family or not, but I became her family. I became not only her spiritual father, but also her family. I became the one she held onto in this most critical moment, and one that brought her hope and assurance. These are some of the things that make me continually believe that God is with me. God is still with me.

Can you tell me about the significance of holy water? I mean, I'm not Catholic, so I don't really know the significance of holy water, its use. Do you perceive it as an instrument or a tool in a way?

The holy water is sacramental, that's what we call it. We have sacraments, and we have sacramentals. The sacraments are just the manifestation of God, a direct manifestation of a way that God infuses grace into people. But sacramentals are tokens of God's love and mercy towards us. And holy water is a sacramental, actually. It's water that is blessed, and it's predicated on the fact that words have power. Words are power. So even in the sacraments, we have the form—the matter and the form. For instance, in the Eucharist, we have the bread and wine as the matter, what you see. Yeah. But the water of consecration is the form. And the blessing of water, we provide the water and we speak to the water, that wherever this water is sprinkled, drive away evil spirits. Let the presence of God be seen and felt by those who use it. So these words transform, as people of faith transform the water. It changes from just water to an instrument of deliverance and healing, and God's presence. That's basically what it is. So where there is suspicion of infestation of demons, people can sprinkle the holy water to ward off those negative forces.

Have you had any experience where you've seen the effectiveness of holy water?

Definitely.

For many years, I've been involved in the deliverance ministry, and the deliverance ministry is more like confrontation with the forces of darkness. And I have encountered situations where I was praying over people and I sprinkled the holy water, and people began to manifest presences that were contrary to the presence of God. And the fact that they feel when you sprinkle the holy water on them, it is like . . . most of them will tell you anyway, it was like pouring fire on them.

And it discomforted them. So what was in them that was hidden began to manifest because of the effect. So as you pray, you sprinkle the holy water. Suddenly, you find out that when the spirit leaves, they're at peace. It doesn't have the same effect on them anymore. I've had those experiences a thousand times.

That many times? Wow. I imagine that's very faith-threatening. You mentioned you've had some near-death experiences. Would you mind sharing with me what those were like?

Sure. I have had two. One was in 2000. I was in Port Harcourt, Rivers State, Nigeria. It was during one summer pastoral experience. Normally, when the seminary was on vacation, the seminarians were sent out for practical experience. So, we were posted to parishes. I was posted to this parish, St. Peter's Parish, in Port Harcourt Diocese. But the parish had other stations that were far from the parish, and the priest had to go to those places still. So when you come for pastoral experience, they position you in various places similar to this. It's important for someone to be in touch with the local parishioners directly.

As part of that ministry, we do visitations. We visit homes, we go to people, we ask what their challenges of faith are. While on assignment there, I had done a normal visitation one day, after the morning prayer, and came back to my lodging in the church around 3:00 in the afternoon. I was tired, particularly more so than was usual for me that day. So I got my Bible from the side of the bed. I was still wearing my cassock.

The door was closed. But then as I just lay on the bed—I had my legs out, of course, just reclining somehow—I saw some beings passing, going through the room, like it was a road. The room was like a road. And these creatures, they looked human, but they were spirits. It was clear. This was not a vision of the night. This was a clear, corporeal vision. I was sealed and passing.

I didn't know what was happening. When I exclaimed, I said, "This is scary, Lord." Then I began to pray, "Lord, Jesus, I am coming to you. Lord, Jesus." I don't know where that came from, but I began to pray that prayer. Suddenly, I saw my spirit come out of my body. I was standing by the bed. I knew this person was me, but the person I was looking at on the bed was also me. So my spirit was able to look at my body on the bed, but my spirit was now praying, "Lord, Jesus, I am coming to you. Lord, Jesus, I am coming to you." As I prayed, I was lifted up, and I saw my hands were lifted, and I was praying the prayer. It all happened like it was two or three minutes, but it was two hours.

"Lord, Jesus, I'm coming to you." I passed the first cloud. Okay? That cloud was so dark. It was so dark that you couldn't see the person next to you. When you talk of the valley of the shadow of death, I understood that later, when I started looking at Psalms 23—but it was so dark. So I went through the dark cloud. The next cloud was so white that even the snow can't compare with it. I went through the white cloud, and the next stage was a smaller cloud. And I understood—there was no language, nobody spoke to you, you just understood. I understood that if I go through this one, I can't come back here.

You just felt it?

I just felt it. I just knew it, that once you go through this one, you can't come back here. I was still praying, "Lord, Jesus, I'm coming to you." As I tried to go through the way I'd gone through all along, I heard a voice, and that voice was like thunder. It was not any language, but it wasn't intimidating, and I wasn't fearful. It was so powerful, from one end of the sky to the other, but I understood what he said. I saw a hand above that cloud, and that hand was bigger than all of Maryland. That voice and the hand came at once. And the voice that thundered, I just understood what he said. He said, "It's not his time." And as soon as that command was given, that hand pushed me back, and I came down. It's like falling from Mount Everest. I came and crashed into my body. You're trying to catch yourself so you don't crash. I came back with force. I got up and sat up on the bed, and I was in the hospital. People were gathered, and people were crying. So, what happened was that in that small church, we had a doctor who had a clinic nearby. So apparently, while I was having this experience, the chairman of the church, who came several times a day to check on me, had come and seen me in an unconscious state.

This is after you were shot, or something different?

No, long before I was shot. I was still a much younger person then.

I'm sorry. I don't mean to interrupt, but did they have an idea what caused it?

No idea. I had just collapsed for no apparent reason. The chairman just came in and saw me on the bed, in my cassock, and could tell something was wrong with me. So, people at the church picked me up and carried me to the hospital. They removed my cassock. When I woke up at the hospital, all I remember was wearing a singlet and boxers, and that was unusual for me.

So you had no idea how you got like that?

No idea. People were so distressed. They were crying, and they were doing CPR, trying to resuscitate me. I sat up, and they held me. They said, "Don't let him stand up by force." And I said, "Leave me, leave me, leave me alone." I got up and I looked around, and there were church members everywhere.

You didn't know what was going on?

No idea. I felt like I was just waking up. So the doctor looked at me, said, "He's okay," and after a brief period of observation, they drove me back to the church. People came, and they were clapping and singing and praising God that I didn't die. But it taught me that death is a beautiful thing.

You weren't afraid anymore.

No. I've never been afraid of death anymore. Never been afraid of death anymore.

Wow. How old were you when you got shot?

I was shot in 2003. I was twenty-nine, and I was going to be thirty.

And before, the experience was three years before this? The other experience?

Yes. I was shot in Abuja, Nigeria, the federal capital territory.

Were you just accosted in the street or somebody tried to rob you?

No, it was in the room. It was in my brother's house. I was on vacation, and I was in the room. It was FCDA Camp, Federal Capital Development Authority Camp. He lived there because he's a civil engineer. He'd lived there for eighteen years, and there'd never been a robbery incident in that place. It was a Sunday night, breaking Monday morning when it happened. I knew something was wrong hours before, but I didn't know what it was. So I told my brother's wife that I wasn't going to eat. When I came back from church, I started fasting. I locked myself up in the room and I prayed all day. I mean, like . . .

You could feel something was coming?

Yes, because I kept crying. Even in church, I was crying. And when I went to the chapel, after mass, I said, "Lord, I don't understand what's going on." I received no vision, no instruction, but I decided to seek God that day because I knew something was wrong somewhere. I didn't know exactly what it was. I didn't know exactly what was coming.

I locked myself in that room and I prayed, literally. I just prayed all day. Then around midnight, after I had gotten a little sleep, I hadn't eaten all day up to that time, and I wanted to do it for three days—I wanted to do three days of fasting with no food and no water. So it was less than twenty-four hours when this thing happened. I got up at midnight, 12:00, and I began to pray. While I was praying, it was about, between 1:45 a.m. and 1:50 that morning, I heard this voice, this very ugly voice say, "Don't move. If you move, we get you down."

I'd never encountered armed robbers. I'd never heard that kind of people before. But I realized in that moment—it wasn't even seconds—a voice spoke to me. It was not audible; I heard the voice clearly within me. He said, "They've come to kill you. Pray." And I still remember what I prayed. I just stood there in the room, and I said, "Father, you can do all things. Save me now," because I'd read the scriptures and I had changed the light to blue, because I prayed the rosary; then I read the Bible, and then I turned off the brighter light. It was dim, but I was just praying there. And this was the last prayer I prayed when the gun went off. It was a moment. It's not even seconds, but an instant.

I prayed that prayer. And in that moment, they broke the door and came in. At first, it was like a dream. But the impact of the gun, because of course, it was coming directly with fire and all that, knocked me up, lifted me, and threw me down. And in that moment, the same voice spoke to me. He said, "If you get up, they will know you are not dead. They will kill you." So I remained on the floor. This was happening in less than seconds. It was more like awareness, and I remained on the floor. Then they stayed, trying to make sure I was dead. They waited about five minutes before they left.

When they left, eventually, I got up and I saw blood all over the place. I went to the hospital and the police came. They couldn't save my eye, but I lived. Up to today, there've been no arrests made. So many questions remain unanswered. But I went through that process, and I can look back now, and I can say comfortably, I think it was probably the final experience I'd had in my long battle with darkness. God used that to show me yet again his supremacy, and it is a very comforting feeling. The lesson that experience taught me makes me who I am today.

If there was a way, and I think there is, that we could prove that spirits, whether good or bad, exist scientifically, do you think it's something that should be done, that science should pursue? Or should it be left to a matter of faith?

Science should pursue such knowledge, such endeavor, because even theology, the study of God, fails to explain certain endeavors. And the Logos is an organized study of God—it's not chaotic. It's not left to everyone's devices. That's what we do. The study of phenomenal reality in philosophy is organized. And there is something that it does—it sets a precedent for further inquiries. And the duty of man is to constantly press into the deeper dimensions of reality, into the deeper understandings of the dimensions of reality. That's the duty of man. All men, by nature, desire to know, according to Aristotle. We don't just know half and half, we want a systematic, more organized form of knowing. That's what should be pursued.

So this study is something that should be pursued so there is something certain for people who don't belong to any religion. Well, they say they don't believe. I don't believe that there are people who don't believe in anything, because some people believe in themselves.

With science, somebody who doesn't believe in God, officially, can still benefit from understanding that there are supra-sensible realities, including *spirits*, that are indeed real. Sure, I might not be seeing it right now, but it doesn't mean it doesn't exist. It's just that I'm not having a direct experience of it, but it exists. So I think this is something that should be pursued because if we understand, like I do, coming from the Igbo African culture, that there is a spiritual traffic between the earthly life and the spiritual realm, we understand the impact that *spirits* have here. Even as a Catholic priest, angels, we invoke the angels. We invoke the ministry of angels. We ask for their aid. We implore God to send your angel to guide us as we go, to send Raphael for healing, to send Michael for warfare. So if we understand

that these beings have been positioned by God for certain assignments, we understand that they must be real.

God's use of angels for service, for assignments, helps us understand that there is an interconnectedness of reality, rather than the
isolationism which the Western world pushes. This is paper, but this
is a book. If we understand that this is paper, but you can make it into
a book, then it makes a difference. I think you're doing something
very worthwhile.

An Interview with a Native American Shaman

It was a perfect-weather afternoon in early spring when I first met Wolf at his home in Santa Fe. I'd been looking forward to speaking with him in person after we'd shared an initial FaceTime pre-interview a few weeks prior. I was convinced his portrait would make a nice addition to a photography exhibit I was completing on traditional healers at the time, and I was curious what his perspective as a Native American healer would be about spirits. My wife accompanied me to the interview, an atypical luxury, for it's one of the only times she's ever been present while I interviewed someone in person. While she usually prefers to remain distant from my interviewing and photography work, I twisted her arm with a promise of dinner at one of our favorite local tapas restaurants after the interview. It was a treat that would end up being a secondary reward, as she found herself as captivated as I by Wolf's wisdom, candor, and calming tone. While my thoughts about the importance of ancient lexicons, cosmology, and Indigenous beliefs had long been well formed by the time of this interview, Wolf's insight on the connection between the term "spirit" and its multiple usages as it pertains to certain Native American peoples was particularly insightful.

Author: *What is a spirit?*

My first thought is, what isn't spirit? [laughing] Most people think of spirit as beings, spiritual beings, outside of ourselves, and yes, that's true. It's the trees, it's the animals, the plants—everyone we meet is spirit, physical and nonphysical. Energy is spirit. The stones have spirit, they have energy, they have a life force, that's why so many people wear stones. People think of spirit as a place where we come from, and the place where we return to, the source of creation, the Creator, Nature, the Source of Life, the Universe, Great Spirit, God, and so on and so forth.

Sometimes when we pass away, some people when they pass away, and sometimes if it's tragic, they get confused and stuck in this dimension, and their spirit doesn't move on. So there are these things that some people call ghosts or spirits. Some of those spirits are beings that have chosen to stay here. They come here to help. They have chosen to help us as teachers and guides.

There's so many answers to that question, but for me, ultimately, it's more about what spirit isn't. Indigenous people, I would say some Indigenous people, maybe, I don't know, all Indigenous people . . . I don't know if there's a word *energy* in some of the old languages or *spirit*; those two words were synonymous with each other. And we know everything has energy. Even these things we think of as inanimate objects have energy—there is a vibration.

Second question: What would the relationship between humanity and spirits be like in the best version of human civilization?

Whew—well, we've lived in that way, and there have been civilizations that have lived in that way. Working in unison, I don't know if working is the right word, living in unison, in understanding that we're surrounded by all these beautiful ones. And calling upon them, honoring our ancestors, honoring spirit, and summoning them for their support and guidance . . . .

Invoking and summoning—if that's the right word, I'm not sure— these helpers, and ancestors, and guides, spirits of the land, for help and guidance. I mean, ultimately, we do live with them. Our language, this language, this patriarchal language has created an illusion of separation. The idea that we don't live in unison with them is an illusion, and the language that we speak doesn't open the door for that.

Living in unison with spirit is part of the evolution, the evolutionary process that we are in the midst of, and the direction in which we are going. It is also a part of the place where we've been. I think of the Mayans, Egyptians, a lot of the Native peoples on this continent and the world, the Atlanteans, Numerians—they had a deep understanding of this relationship, and their language reflects that deep relationship.

In our traditions, we're taught to talk to plants, and to animals. We don't just go out and cut the herbs and the medicines without talking to the spirit of that plant and making an offering, without stating our intention or our prayer. Dances are done to call upon the spirit of the deer, the elk, the buffalo before the hunt to honor their spirit, to show our gratitude and reverence, knowing that we cannot live without those plants and those animals, without the sacrifices that they make of their life so that we can live. We honor that life, that spirit. I think that the deepening of a relationship with humanity and spirit opens the doors to infinite possibilities, and if we can imagine it, we can create it. This is also a deepening relationship with our own self, our higher self, our whole self, which includes all that is.

When we are in joy, we are closest to Great Spirit, and we can feel that relationship, that closeness. It is much more accessible when we are in joy. If we meet our pain and discomfort with joy, there's no overwhelm, and there's a greater opportunity for learning and growth—it's so much easier and faster.

Do you have a business where you provide healing for others?

It's so difficult to refer to what I do as a business. It's so difficult to make one's spirituality into a business. You know? I don't feel comfortable describing it in that way. I do private sessions, healing work, counseling, and stuff like that. Some of the work I do is hands-on and bodywork, and amazing things happen. Yes, I do charge a rate for that session. At the same time, I'm really flexible with that rate because I don't think healing or medicine is limited to those who can afford it. That's the problem with our society. I have a mortgage and I have to pay the bills, so it's difficult or challenging. It's an interesting challenge to find that middle road, you know?

I frequently conduct various ceremonies, and yes, people offer donations for ceremonies. We don't ever ask for a specific amount. There are no fees for ceremonies. It's just all done by donation and whatever anybody is willing or able to offer as gratitude, as an exchange, for that medicine.

I understand completely. You look at some places like India or Cuba where everybody, for the most part, most of the people are impoverished. You look at what is needed to sustain life, versus America. It's much more expensive. You have to have balance. What you have is valuable to many, and you still have to live in society. You have to balance those needs. Do you mind if I just go through some basic background stuff?

Mm-hmm [affirmative].

Where were you raised?

Denver, Colorado.

Okay. And you say you're Lakota. Is Lakota heavy in Denver?

No, I'm Spanish, Navajo, and Ute. I adopted Lakota. The traditions I follow are based on Lakota tradition. You could say I'm Lakota, in a traditional sense.

Okay. How did you become exposed to your spiritual tradition?

That's a big question. Okay, my spiritual tradition. . . . My earliest memory is seeing a ghost coming down the hallway toward my bedroom as a little boy, maybe four or five. When I saw that ghost, I got excited, because I thought that Casper was coming to see me. I was so happy to see this spirit coming. It was an apparition, like a whitish fog. But it had form, shape, kind of a human form. As it was moving, I got really happy. Those kinds of experiences were happening almost daily.

How old were you?

Four or five. We lived in that house until I was thirteen. I would say almost daily, I could see them. I could feel them. I was very aware of their presence. I was being raised Catholic, and my mother and my elders were teaching us that there was no such thing as ghosts. I was thinking, "Oh well, they must not know or they must not see." You know? But I was aware that there was something there.

I didn't believe them when they said there's no such thing as ghosts. I was like, "Yes there is." One day they [the spirits] were making a lot of noise, and I was trying to watch *Star Trek*. I was about seven years old. I got so upset because they were making so much noise and being distracting. I just wanted to watch *Star Trek*. I was telling them to stop making so much noise and to shut up. In my anger and confusion, I spoke out, "I don't even know who you are, or where you come from, or what you are. My mom says that there's no such thing as ghosts, but I know you're here. Who are you?"

I heard a very clear voice say, "The answers to your questions lie with your Indigenous ancestors." That turned on a light around my wanting to know the spiritual practices of my ancestors, my Native ancestors. My mom didn't know anything about those things. At the age of thirteen, I met my first healer and my first teacher. She was a psychic healer. She was teaching me. She's the first person that said to me, "Oh, the spirits love you, and you have a gift."

This was in Denver?

Yeah. She was the first person to tell me that I had a gift. I was so excited. I wanted to learn more about that and how to utilize that gift and how to work with that gift. I started studying about healing, communication, and all kinds of esoteric practices.

What was her tradition?

She didn't have a tradition, per se. She just referred to herself as a psychic healer.

Nobody in your family before her had any experience interacting with these beings?

Exactly. At least nobody that I was in touch with. Looking back, I remember an uncle that worked with herbs and made medicine from herbs. Not just an uncle, but also a grandmother, and a great-grandmother. I'm sure there were other elders, great-uncles, and aunts that did those things, but nobody I was close enough to that had the kind of knowledge or understanding that I was seeking at a young age.

Did she teach you anything? Were you able to make some progress in some way?

Oh, yeah. For sure. She taught me communication with spirits and healing, doing healing work. All kinds of things. Meditation and seeing auras, telekinesis, moving things with your mind. You know, things that were just popular in those years. Telekinesis, I haven't practiced anything like that in so many years, as I don't see the point of it. It's not necessary for healing.

It wasn't until I was twenty-nine that I went to my first Native ceremony. It was a time in my life when I was living in Manhattan and I'd decided to end my life. I'd been doing drugs and alcohol, and I went into that dark, bottomless pit. It felt like there was no escape. It was there, when I was in that place emotionally, that I was invited to my first Native gathering. It was a four-day gathering in upstate New York.

Who was that with?

With a medicine man.

Was he Lakota, or . . . ?

No, he was Mi'kmaq. Mi'kmaq, they're from Canada. He was holding a four-day gathering, and I went to that gathering. I was told to be drug- and alcohol-free for four days before going, and four days after. I hadn't been sober for four days in fifteen years.

Wow.

Yeah. That was a big thing.

You must have felt really compelled to be a part of that ceremony.

Oh, there was no question that I was going to do it. I'd been waiting for an opportunity to go to that kind of ceremony my whole life, since I was a little boy, since that voice said the answers to my questions were with my Indigenous ancestors. That was something that motivated me for many, many years. But by the time that invitation appeared, it wasn't about knowing about the spiritual traditions of my ancestors, it was more about saving my life. Those four days totally changed my life. When I left that gathering, I never did drugs or alcohol again ever.

I started studying shiatsu, bodywork, and healing. I worked with a Japanese master. I found another elder that would become my teacher. He was Lakota. That was about a year after I was sober.

What was it about the experience that was transformative for you?

Everything. It felt like being home. I just felt like I had come home. Being a misfit my whole life, home was a feeling I never had. Of course, I felt at home with my family. I come from a very, very loving Hispanic family and upbringing. I'm very close with my cousins. But there was always a part of me that didn't feel like I fit in. Maybe that was the part of me that was gay. Maybe that was the part of me that was different. That's why I always felt like a misfit. Those ceremonies, those first ceremonies at that sweat lodge, I really felt like that was where I belonged.

So for the ceremony, he did a sweat lodge? Was that just that part of it?
He did four sweats in four days.
During the sweat lodge, did you interact with some ancestors?
Yes.
Can you remember what that was like the first time? Can you tell me a little bit about what you experienced?
Like I said, it was like coming home. I started seeing spirits again that I hadn't seen since I was young. They welcomed me. They took me. It was like a reunion. It felt like a family reunion with the spirits. I was so happy to be home. When I left that gathering, I didn't have any desire to go back to my old ways. I didn't have any interest in putting myself back into that bottomless pit with drugs, alcohol, and depression.

Did you have an experience, as part of that, that showed you what you should do?

Not yet, not at that point. What I should do? I knew what I wanted to do. It's not a matter of should. I wanted to dedicate my life to these ways. I wanted to learn as much as I could. It had nothing to do with if I should. It was passion.

Yeah. I know exactly what you mean. I've felt that a few different times in my life. It's a hard thing to explain to others. I can't really explain it anyway. It's just something you know you have to do, kind of, right? That's what you felt?

Yes.

Okay. Is the Lakota elder the one who showed you how to go perform a sweat lodge?

Well, when I became initiated with him, for him it was just about dedicating my life to these ways. I had found home. I had come home, and I wanted to stay there. It wasn't about wanting to become a ceremonial leader or running sweat lodges. It wasn't about that. It was about participating and doing what I felt was my calling. After doing that for about eight years or so, he said it was time for me to start leading the ceremonies. I was like, "Well, I didn't know that I was in this to start leading." But that's how it went. You know.

Is the initiation that you went through something that you can discuss? Did you make any promises as a part of it?

Oh, for sure. I dedicated my life to taking care of the people. I dedicated my life to taking care of these traditions and taking care, as best I can, of the people and myself, and to bringing more beauty into the world, to trusting in these ways and these spirits. That's just to name a few. There's a few others, but I don't feel comfortable sharing.

Sure, yeah. I don't want you to share anything that you're uncomfortable sharing. As part of the tradition that you're initiated in, do you promise to not use the gift to hurt people? Is it just for healing?

That's kind of a moot question. Why would anyone? I don't know why anyone would. Of course, I don't hurt anyone. My life is dedicated to helping lift the people, helping to bring medicine and healing, to relieve suffering and pain.

Now, when we talked before, you mentioned that over the years, you have interacted with healers from various traditions. I think you mentioned Yoruba and you mentioned santeros. . . . I can't think of the others you mentioned. I was wondering if you could speak about those experiences, meeting healers from other traditions, how that affected you, maybe what similarities you found.

Sure. I've had the gift, the huge blessing of meeting many elders, many of whom I'm still involved with, but not all. I've met people of all kinds of traditions, not just Indigenous and Native American, but from all kinds and different walks of life. My focus is on Indigenous traditions. Indigenous means "of the land." My interest has always been in not necessarily Native American traditions, but in Indigenous or shamanic kinds of traditions. I've spent a lot of time with some Yoruba elders. My husband is initiated in those traditions. I've spent a lot of time with people from Shoshone, Lakota, Apache . . . yeah, too many to list . . . Blackfoot, too many to name, really. All various elders and from different places. I've spent time in Mexico, with elders in Mexico . . . and Peru, with elders in Peru.

How did you find these people?

Just through life. I wasn't necessarily looking for them. They just came. I met a lot of elders in ceremonies, various ceremonies that I would attend, or who happened to be a part of them.

Did you ever feel like your spirit guides were guiding you to interact with different people for certain reasons?

Undoubtedly.

What were some of the reasons you perceived that you met certain people? To help them in their way? For them to help you in your way?

That's true of all we meet. I don't believe in coincidence. There's no such thing as coincidence. There's no such thing. It's all part of a greater plan that we're not fully aware of. Some people are drawn together, and their energies feel cohesive.

Yeah. Sometimes friendship is unspoken. You just feel it.

Right, exactly. Sometimes? Always.

Did you go through the process of making some sort of sacred space?

Yeah, of course. That's one of the first things you do. You know, you build an altar or create your sacred objects.

Were you guided in that process?

Oh yeah, that's part of the training. My teachers were showing us how to make our altars, our prayer ties, make our *chanupas*, our secret pipes, make all of our medicine things, our medicine shields, the clothes we wear for our ceremony, our regalia. Everything we make ourselves—our drums, our rattles, all of our medicine things.

Of course, when you're young and you're doing it for the first time, you've got to have someone there showing you how to do it. It must be done in a good way, in a prayerful way, in a sacred way—all those things.

Okay. After you started constructing it, what was the consequence? Did you feel like your intentions were supported more, that the things you were hoping to change, you were able to do things more effectively?

Absolutely, yeah. "Consequence," that's an interesting choice of word. Everything we do . . . when we do these things, we're doing it to have an effect. We're doing it for a reason. We don't pray because we don't want anything to change. We pray because we want something to change. Everything we do has a consequence. You can say good or bad, light or shadow. The more conscious we become, the more we become aware that every thought, and every word spoken or unspoken, is going to have an effect. If negative things keep happening in your life, then you might want to ask yourself, "What am I doing that keeps causing these experiences, and what can I do differently to put out energy that is more supportive or joyful?" This is why we must keep looking at our shadows. That is really important. If we're not looking at our shadows, we cannot grow. It's not just our shadows, but our light, too.

So many people are so focused on their insecurities, or fears, or their anxieties, and they aren't looking at the beauty that they have.

So, they are ruled by those fears, anxieties, and insecurities, and not aware that they are beautiful. If we're putting a lot of our energy into our insecurities, the universe is going to reflect all those things back to us because the universe wants us to stop it. The universe wants us to change. The universe is trying to show us ourselves.

One of the most beautiful things I've learned or experienced in ceremonies is going into the ceremony and realizing, "Wow, I have a lot of beautiful things. I have generosity, I have laughter, joy, compassion, I have the desire to help. . . . I care." Seeing those things magnified also, which is our light and our beauty. Seeing those things magnified, it's easier to see those shadows and say, "Okay, I see how you've been moving to help me realize who I really am. Not that I am insecure, but that insecurity is actually trying to show me how to be more secure." Meet those insecurities and focus our heart, mind, and energy on what we can do to not have those insecurities.

Many of the individuals I've talked with describe having clairvoyance, and it manifests in different ways. Some people have experienced unique dreams. Others can intuit others' emotions and aspects of people's character . . . what physical ailments a person may face, relationship issues . . . without even having any . . . just being around the person. Is that something you've experienced? Do you mind discussing ways in which you've experienced things like that?

Body language. Intuition. . . . How do you describe intuition? Intuition is a feeling. Intuition is an inner knowing, and it's about learning how to trust that inner knowing but never to make assumptions. It's about having a sense of what somebody might be. I'm using insecurity as a metaphor, so I'll just stay with that. Somebody might be feeling insecure in a particular situation, and you can feel that because you have that memory. I have that experience also, so I have a feeling of what this person has gone through and I've been there. But I'm not going to assume that this is what you are feeling or what you are doing.

The way I approach it is that I ask them a question: "Are you feeling insecure right now? Well, this is what's helped me in the past."

I don't like to tell people. I don't think people like to be told, "This is what you're feeling. This is what you're doing. This is why you're doing it." It's so invasive. That's not cool. It's gentler, it's more motherly and nurturing to say, "I have a sense that you're hurting. Tell me what's happening with you." I might know what's going on with them, but I'd rather ask them if they want to share.

Do you feel like your guardians direct you as to the course of helping a person?

Sure. They are the ones doing the work. I'm not actually doing any healing work, I'm just the vessel. I'm mostly doing anything I can to get out of the way so that the medicine can come through. Back to the word you said, clairvoyance. If I get out of the way, then that clairvoyance or intuition from the spirits can come through. They're the ones doing the work.

How does sage assist you? I've burned sage. My wife in her Cuban tradition has used sage in different ways. I know how it makes me feel. Is there a specific reason why Native Americans use sage? How does it assist you?

I actually use cedar more than I use sage, but I use sage too. I use these different medicine plants in different ways. Each one has its own kind of medicine. I use sage daily, but I also use cedar often. Cedar is for cleansing. It's all for cleansing, these two plants we are talking about. We're talking about cleansing. Sage is more physical—cleaning our physical body, our house, our physical space, or our sacred objects. Cedar's a little more for mental and emotional cleansing.

Have you ever used any other assistive herbs, like something similar to peyote or ayahuasca? Are you familiar with these things and their purpose? How would you describe the appropriate use of these things?

Yes, I have experience with some of these teacher plants, as we call them. I've only used them in a traditional setting, in a ceremonial setting.

Not for recreational use?

Oh, when I was a kid. Like I said, I went into that dark, bottomless pit back in those years. Of course, I was experimenting. I was addicted to marijuana and alcohol, but I'd also play around with other drugs. It was nothing to do heavy mushrooms, as an example. Mushrooms are a plant, a teacher. In those days, I didn't know that it was a teacher. I just thought it was a way of getting high and hallucinating. I didn't really approach it with the respect and reverence that I've since then learned.

We approach these plants with respect and reverence. They teach us so much. They show us so much. They carry an incredible wisdom and consciousness. They're so intelligent, these plants. That's why Indigenous people, some Indigenous people, have used these teacher plants for many, many, countless generations, including sage and cedar. We use these plants because of their intelligence, because of their wisdom. I feel that the plants are so much wiser than we know. They are so much more intelligent than we are. They have such wisdom that we can learn from.

Like, "Take a look at the lily," said Jesus, and he spent much of his time teaching his disciples in the Garden of Gethsemane. Could you tell a difference in the use of the plants in ceremony versus recreationally?

Yes. It was more respectful. In ritual, it would always show me something or reveal something to me I needed to see.

Do you feel like, over the years, you've become more skilled in your ability to help others? What would you say are ways in which you have grown?

That question can be applied to anybody or anything. You've been cooking for a year or you've been cooking for twenty years. If you've been cooking for twenty years, you're going to definitely have perfected your art. No matter what it is—carpentry, singing, painting, or counseling. Hopefully, you've grown. Whatever it is you're doing, you've grown. If you haven't, maybe go on and do something else if your heart's not in it. By just practicing, we learn. We learn more from practicing than reading. You can't really learn a lot in these ways from just reading. I haven't actually read any books in years.

Are you familiar with, in Cuba, it's described as "the Bad Eye," but in other cultures there are similar names. . . . Are you familiar with this?

No.

Okay, "the Bad Eye" is a bad energy where thoughts of jealousy or people's bad intentions toward you can effect general malaise or cause you to have problems.

Who doesn't know that? Everybody knows that. Every time you say "f-you" to somebody, you're doing that. Every time you say a curse at somebody, you're putting them down. Every time you send a negative thought to somebody, you're doing what you call a "bad eye." Every time you're driving down the road and you get cut off and you curse at that driver, you're sending out bad juju. You're sending out negative energy.

Is it common for Lakota or other Indigenous practices to incorporate water in their sacred space, like a glass of water?

You can't do a sacred ceremony without all the elements: fire, water, wind, and earth. They all have to be present. That's also true with a sweat lodge. We put the stones in the fire. The fire is an essential part of the altar. You heat those stones in the fire. We bring them into the lodge. We put on the water. The lodge itself is on the earth. It's on the ground. With the help of the water, we release the breath, right?

Those are all four elements and all four directions. In the ways that I've been taught in ceremony, in our way and tradition, you have to have all four elements present, which incorporates all four directions, which is the whole of the universe. All of these things give life. Mother Earth, we couldn't live without her. The wind, the breath of life, we can't live without that. The fire? We can't live without that. The water? We can't live without that. These four simple, basic elements, they are alive. They are beings. They are alive. We honor all of those beings in every ceremony.

During the sweat lodge, is it customary to feed the spirits in some way?

Mm-hmm [affirmative]. Putting food up on the altar for them. We feed the ancestors almost daily. We feed the ancestors constantly. Also, when we go in the sweat lodge, by singing these sacred songs, we're calling them in. It helps to strengthen them. It strengthens us by putting in our time and energy as a sacrifice. It's an offering by opening our heart and speaking from our heart about our pain and the things we want to get rid of. That also helps feed them, especially by expressing our gratitude. That definitely helps feed the spirits, offering our gratitude.

You know, to answer your question, what isn't an offering? It's all an offering. When we become aware of that, it's all an offering.

Is animal sacrifice common in many Native American traditions? Like, blood?

I was talking about the intelligence of the plants and the trees. The animals have great intelligence. The eagle doesn't ask how to be an eagle, or what it means to be an eagle. An eagle just is. A tree doesn't ask how to be a tree. The rose doesn't ask how to be a rose. The wolf doesn't ask how to be a wolf. Humans are the only ones that ask how to be human. We're the only ones that have forgotten our original instructions. We've forgotten. Some people remember the original instructions and live by them, which is to take care of each other, and to take care of Mother Earth, the water, and the fire, and the wind. We're caretakers and givers. We're caregivers. We come here to give care because the plants and the animals, they don't need us. They would be much better off without us here, right? But the plants and the animals, they know that in order for us to live, they make a sacrifice of their life so that we can live. Every time you eat, whatever you're eating, whether it's plant or animal, something died so that you can live. Every time you eat, every bite, someone made a sacrifice of that animal's life. That animal made a sacrifice for you.

So, is there animal sacrifice? How would we live if there wasn't? We have to ask the animal to sacrifice their life so that we can live, and the plants. They know that. They're aware of that. That's why every time I have a meal, I offer food to my ancestors, but I'm also making an offering. I always say thank you to the plants and the animals for their life because I know that I couldn't live without them.

When you have students . . . I imagine you've had apprentices through-out your life, just like you were once an apprentice. What are the most important things you try to impart to your students?

Trust. Wisdom. Trust is so important. We have to trust ourselves. We have to trust Mother Earth. We have to trust life. We have to trust each other. You can't have a healthy relationship without trust, right? That's what life is about. It's about relationships. I feel that's what it's all about, anyway. Every ceremony we do, everything we do, is about building trust. It's about building relationships. A relationship with Mother Earth, relationships with the plants and animals, the wind and the fire. The understanding that we're related to the sun and the moon and the stars. We're related to all people. Black, red, yellow, or white, we're all related. That's what we say when we finish our prayer. We always say *"Mitakuye Oyasin,"* which means "all our relations." When we say that, we understand that we have a place in the web and that everything has its place in the web. When we say *"Mitakuye Oyasin,"* we're saying that we all belong, that I belong, and that we're all related.

We all want to be in a healthy relationship. When I use this word *prayer*, it's not about worshiping the sun, or the fire, or the moon, or the ground, or the animals, or the plants. We move to talk to the spirits of the animals and the plants. We move to talk to the fire, or the water, or the wind, or Mother Earth. We move to talk to the stars, the sun, and the moon, to all aspects of nature, because you can't have a relationship without communication. The only way to be in a relationship is to communicate.

To have a healthy relationship is to communicate, to open up, to share ourselves. *Prayer* is a word that I use, but not in the traditional definition of worship. The way I define prayer, it's about communication. It's about wanting to be in communication, in a relationship. In a healthy relationship, we are open with one another. The way I was taught is that we go into a ceremony to talk to our ancestors. We go into a ceremony to be in a healthy relationship.

The elder says, "Hey, say a prayer for us." We do something inside so that our words can come from our heart, from that sacred place. We don't use curse words. In old languages, there were no curse words. People didn't curse each other because there was no concept of that. When we go into the lodge or into the ceremony or whatever, when we sit down to talk to our ancestors, our words come in a respectful and reverent way, in a humble way. We have to be humble, we must humble ourselves, right? But it doesn't do any good to do that in the ceremony, then go out into life and curse the driver that cuts you off in traffic. It's got to be all the time.

When *you* go home, when *we* go home to meet our creator, to meet Great Spirit, our home on the other side, Great Spirit doesn't ask, "How many times did you go to ceremony? How often did you pray?" Great Spirit says, "How did you live? How was your life?" That's the real ceremony. Every word you speak with your life, that's your prayer. Those are the real prayers. Those are the ones that matter the most. If we're not treating our friends, family, or lovers with love, respect, and care, what's the point? We have to treat ourselves with love, respect, and care so that we can treat each other with respect and care.

Part of the reason why I do what I'm doing is because I feel one of the problems with society is that it seems the partnership between humanity and the unseen world is not in harmony. We must be in harmony. If you don't even acknowledge that these beings exist, then how can there be harmony? You see it in the way we treat the Earth, and the way we treat each other. Sometimes I, much like you, feel impelled to do the work I'm doing.

Do you think that scientific proof would be helpful to humanity? For example, you look at this water and you study it, and you prove that these entities are real—would that be significant?

That's what metaphysics is all about. Metaphysics is a form of science that is trying to validate the things that shamanic people have known forever. I think it's less important for us to prove spirits to others. It's more important for us to prove ourselves to the spirits.

On one level, on a certain level—perhaps level is the wrong word. It's an English word, but it's the only word that I . . . *layers* is a better word. There's so many layers. There's so many layers and levels perhaps. On one level, I agree with you 100 percent, and yet, if we are really in deep trust of the whole picture, then we have to trust that everything is for a reason. Think of all the racism coming up, and that has come up in the past few years. Thank you, Donald Trump, for bringing all that up for us to look at it. The beautiful thing about it is that it's showing us how compassionate people really are about racism, and how much people really don't like it and want to change it. It's exposed something, right?

There's a reason behind it. You can hate Donald Trump, or you can hate White supremacy, whatever. Some people hate Black Lives Matter. Hate whatever you want, but that's not going to solve any problems. That's not going to resolve anything. The only thing that's going to bring resolve is to care, to be caring. I remember one day seeing a man that was wearing one of those . . . it's called a Police Lives flag, a Police Lives Matter flag. I'd never seen one. When I had seen it, I was like, "What is that?" American flag, black and white, or American flag with just blue stripes. I thought, "What is that?" Somebody said, "It means Police Lives Matter." Does that mean that they're not supportive of Black Lives Matter, or that they're Trump supporters? They said that's basically what that means. My first impulse was to be like, "Oh, I want to go away from that person. I don't want to be near that person. I want to go the opposite direction."

But then, when I recognized that . . . I realized that if I do that, then I'll be creating division. My frequent prayer is about creating unity. The medicine wheel of life is black, red, yellow, and white. It creates the wheel. It's the whole. We're all related, all my relations. When I realized that, I realized one of the beautiful things this past year has shown me is the subtle ways in which I've created division and how I judge. If I see somebody that I might think is a White supremacist or a White Republican, wow, how judgmental I am.

Maybe they are White supremacists. Maybe they are Trump supporters, or whatever. How can I be creating more unity if I don't interact with this person?

What I've learned is to try and find a way to reach the heart of that person, knowing that we all have more in common with each other than we have different. Maybe our skin is different. Maybe our beliefs are different. Maybe how we have sex is different, but our hearts are all the same. We all have the same love, desire, fear. What we have in common is much bigger than what we have that's different. I challenged myself to go. I want to find the place in myself that can talk to the heart of that person, because that's a person. That's a relative.

When I became aware at a very young age, young on this medicine path, I became aware that there are shamans out there judging shamans. There are medicine people judging medicine people. I was so confused by that. I thought that we were all here to talk to Great Spirit. Our ancestors were all practicing the same tradition. Why are those people at that sweat lodge judging people over at that sweat lodge? Oh, because they think they are doing it the right way, and the others aren't doing it the right way. It's that Christian mentality that has been transferred onto the patriarchal mentality: there's just one way, or the right way, or the best way, and if we're doing it the right way, then you must be doing it the wrong way. That kind of mentality doesn't create unity. It creates division.

I've seen it in the santero world, for sure. For sure, there's no place I haven't seen it. Santería, Native American, the yoga world—all of these yoga teachers are out there saying, "I did it this way and my teacher did it this way." That's Christianity, you know, but we don't all have to practice the same spiritual practice. It doesn't matter what you're practicing. If you're a singer or a rock star, maybe you're judging the next rock star because you think you're doing it a better way. But really, the best ones are not there to be the best. They're in there because they want everyone to win. There's no competition. I want you to win. I want you to thrive. I want you to get it. That's the path of love. There's no competition. The patriarchal way is just warriors fighting.

These traditions that I follow come from matriarchal societies. True warriors train so that they can protect if the enemy's coming or so they can go on the hunt. They train so they can kill a buffalo to feed the tribe. They all train to be the best, but they're not training to compete against one another. They're training so that they can be one band of warriors together, united, strong in mind, body, and spirit. Strong in unity together. There is no competition. Maybe they played games or there was competition, but it wasn't about who could be the best. It wasn't about belittling the loser, but it was about helping everyone to become better.

Maybe this is a human thing. I don't know if it's a human thing. We're talking about spirituality. We're talking about Cuba or New York, wherever. We're talking about these people that are practicing bad juju. What I've learned is that if you're afraid that somebody is sending bad juju, and you're open to being hurt from fear, we have to be very, very careful with that. We have to be very careful with our fear.

Love is the greatest medicine there is. Great spirit, Encantada, God, Buddha, Allah, Jesus, whoever you talk to. All the great teachers symbolize love, right? That's what it is. It's about love. If somebody comes to me and tells me so-and-so is sending bad juju to me, I say, "Oh, wow." I pray for that person. I pray for their heart. I pray for their soul. Good, I'm glad that they're sending it to me, because it isn't going to affect me. I'd rather them send it to me than somebody else, who's going to get caught up in that web. Save some other soul from that bad juju. But who's not sending bad juju in this society? Again, the metaphor of being cut off, bad road rage—that's bad juju. Every time you curse at somebody.

Shamans are people. They're human. They get angry. They get jealous. They also do good work. In the moments that they're doing good work, they're doing good work. In the moments that they're letting their ego, their wounds, or their fears get the best of them, those habits aren't serving them. I try to stay dedicated to growing. You can see the ones that are making mistakes. Pray for them. Forgive them. Love them. But if they keep making the same mistake over and over and over for so many years, they're going to hurt themselves more than they're going to hurt anyone else. Continue praying for them. Continue praying that they wake up. It's not just "they," it's me too. I pray that I'm more awake tomorrow than today. I want to wake up. Who doesn't want to wake up? I want to wake up. I don't want to walk around with my fears, suffering, insecurities, and anxieties. I don't want to walk around with that every single day, all day. That's not going to help anything or myself. We've got to take care of ourselves. The more we do that, the more we take care of ourselves, the more we see that it is futile, it is fruitless—that's a better word—it's fruitless to judge anyone for anything.

That's a good way of seeing things, a healthy way. It helps to see things this way.

Are there any prophecies that you get encouragement from? Can you talk a little bit about what encourages you from a prophecy standpoint?

There's so many prophecies I've heard from so many people, for so many years. For as long as I've been doing these things, for close to thirty years, I've been hearing about prophecies . . . Lakota prophecies, Zulu prophecies, Yoruba prophecies, Lion prophecies, Shoshoni prophecies, Hopi prophecies, all these different peoples and all their prophecies. The Hindu, Indian, Christian . . . the list goes on and on because everyone has their prophecies. As I listen to all of these prophecies, obviously, I've spent more of my time invested in listening to prophecies of Indigenous people because that is closer to the Earth and feels more grounded. It resonates more with my soul.

But as I listen to all the prophecies, I take the common themes. I take that they're all saying the same thing, right? They're all saying the same thing. Whether you're talking about prophecies or what spirituality is all about, they're all talking about unconditional love.

They're all talking about how important our words are and to pay attention to our words and our language. It's so important, spoken or unspoken. These teachings are universal.

As I listened to prophecies, a year ago, this massive spirit came and wrapped around the Earth so rapidly. I was like, "Oh, my God. This is the prophecies coming into manifestation." I recognized it instantly: "I see what this is about. This is about rites of passage. This is about initiation on a global scale. This is about change and entering into the next phase of human evolution. This is about all the things that the prophecies have been teaching us all this time. Wow, this is fantastic! I didn't realize I was going to be alive to witness it. I'm alive not only to witness it, I'm alive," I said. Then the spirits said, "You aren't here to witness this. You've got work to do." We have a part. We have a role. We're here to do something.

An Espiritista Shares His Work

There have been many times in the past five years when I've been astounded by the forthright candor of those sharing their experiences with me, and the account below is perhaps the most outstanding example of such. Here, the common pattern many practitioners of spiritism have of caring for unseen entities and then using them to accomplish various works is spoken of in detail. Those familiar with the works of shamans, Wiccan, and other similar practitioners will find immediate resonance with this account, as its obvious authenticity jumps off the page.

Demons don't like the light. When I work with them, I can only work by candlelight, and usually with just one candle. If the electricity is on in the house, it'll go out until the demon leaves the home. I have worked with countless spirits, both good and bad, for many years. But I only do what is good. I do not harm anyone. Even bad spirits can be controlled so that they can be used to heal and help people.

To do the work I do, I must use the bones of dead humans. There is an old, abandoned cemetery known as the Cemetery of the Pirates near my home. It rests on the top of a hill near the ocean where the water is very still and the breeze is strong. When I go there, I feel an overwhelming sense of tranquility. There are no living people there, and all you hear are the sounds of the breeze and ocean waves.

It's there I find the bones I need. When I gather the bones, I do a ritual and say prayers to invoke the spirits to come out. Vortexes of wind form all around the cemetery, and dozens, sometimes hundreds, of spirits rise from the ground. My son, though gifted with the ability to see the spirits, does not do the work I do. The first time he went with me to the cemetery, he nearly peed himself out of fear of what he saw. His dog was with us and started barking as he saw the spirits rise. Both dogs and horses can see spirits. My son said he wasn't sure if he should get inside a grave, throw himself off the cliff into the water, or just leave running because he was so scared.

When the spirits rise from the ground, they have flesh and bones and look just like people. After they rise from the ground, they tell me if they want to come with me or not, and I place the appropriate bones in a ceramic pot and cover it with a lid. I then make an agreement with the spirit. I vow that if they go with me, I will feed them and take care of them in the way they deserve with blood, animals, rum, and other things they may want. In exchange for my care of them, it is understood they will protect me and help me in my work of healing and cleansing. The spirit knows from that point on, it will no longer be lost in the world, and by my feeding it, it will have peace and light again.

When I take the ceramic pot to my home, I place it in a small room under my house. I must feed the spirits right away with an initiation ceremony that must be done on the ground. In that small room, I sacrifice an animal, usually a chicken, a bird, a lizard, or a snake, and then place it inside the ceramic pot. Within a day or so, nothing will be left inside the pot except feathers or small pieces of skin. Only when a spirit is fed in this way will it be able to protect me and assist me in the work I do. The spirits can only protect me if they are strong, and the blood from animals gives them strength.

I am an espiritista. I can talk with the spirits the same way I talk to humans. They are as real as you and I are. Much of the healing and cleansing work I do requires preparation. Although I will often obtain the rum, candles, chickens, and other things needed in the days before a ritual, circumstances often change when the spirit sees the person needing assistance. For example, perhaps we planned to do a ritual with a chicken, but when the spirit sees the person needing assistance, they may ask for a lizard or goat instead. I will often help guide the person into transition with the spirit, and once the spirit is inside the person, it will know everything and anything that was done to the person—who did it, if it was from witchcraft for example, and precisely what to do to remove the work.

I often work with spirits that are very strong in the work of healing. If it is a man who is being healed by a spirit, he must place his hand on his head. This serves to protect his mind and also his fertility. When a woman is healed, she must place her hands on both her head, which likewise serves to protect her mind and fertility, and her breasts, which are the parts of her body that deteriorate the fastest. Both women and men must turn their backs during the healing ritual, or they will see things that no one is permitted to see.

When I was a young boy, I would see the shapes of shadows and people walking that I knew weren't really alive. I knew they weren't real, but I couldn't explain this until I realized that, like my grandmother, I was given the gift. My grandmother was a santera, and she had an altar in the house. For years, I would watch her work, but as she grew older, I began to be able to interact with her altar. At first, when I was about fifteen, it would be simple things, like, I could sense when a friend was going to call me or come by the house. There would be other friends already at my house, and I would tell them, "Watch this. So-and-so is going to call me." Then the phone would ring, and it would be that person. Later, as I grew stronger, I was able to summon people to my home. I would stand in front of the altar and ask the saints to bring someone to my home . . . and in a short time, they would come. This amazed my friends, but not me so much. I already knew what the spirits could do. I had seen my grandmother help many people. Sometimes the things she did amazed me, but as I grew to know the saints, I realized that what she was doing wasn't coming from her, but from the spirits.

My grandmother died when I was eighteen. When she died, the spirits would bother me every day. I could see them, and they would keep me awake at night. When I tried to sleep, they would talk to me or touch me. This happened for weeks, and it happened so much that I got sick of it. One day I went to the beach and took the altar and all the things my grandmother had on it. I threw everything into the sea. When I got home, the spirits were gone. But it wasn't over. They began knocking on my door and asking to come back in. There was one spirit who was particularly menacing. He was louder and more aggressive and wasn't one of the ones my grandmother worked with. I asked the other spirits to send him away, but they said that they couldn't. They said they needed him because he was older and stronger and knew more than they did.

For weeks, the spirits kept knocking on my door and calling out to me. But I was determined. I didn't want to have to deal with making an altar, taking care of the saints, or letting anything into my body. One night, however, my grandmother appeared to me and told me to work with the spirits. She assured me that they would not hurt me and that I would be used to help many people with my gift of healing. I reluctantly agreed, but with one condition. I told the spirits I would take care of them with cigars, rum, and blood, but nothing was going to go into my body.

I built an altar and gave the spirits what they asked for. They respected my wish. I do not need to go into transition to accomplish my work. I simply talk to the spirits. I have helped individuals with all sorts of problems, but my favorite is women with fertility and pregnancy problems. I have seen many women who desperately wanted children but, for one reason or another, were plagued with multiple miscarriages or infertility. Sometimes the answer is simple. The spirits instruct me on how to prepare a fragrance or necklace or herbal mixture. At other times, the solution is more complicated, and blood or bones are needed.

As I've grown older, my abilities have grown stronger. When I walk down the street, I see many people. I only need to look at them, and I can sense what problems they are having and if they are doing evil things. This can be overwhelming. It can be challenging to know intimate things about people just by seeing them. Now I typically walk with my head down, so I don't have to experience that strong empathy when I go out. I like being with friends and family, and I no longer see caring for my saints as a burden. I don't let my healing work keep me from doing things I enjoy.

I only work with the good. I don't use my gifts to hurt people, but I have been selfish in the past. When my mother was ill with cancer, I was desperate to save her life. She was in the ICU in the hospital, and the doctors had given her only a few days to live. I had been speaking with the spirits, and I knew the doctors were correct. I asked a powerful saint about what could be done. He told me he could save her life, but that there would be a cost. He told me he would come with me to the hospital, and in exchange for giving more years to my mother, he would have to take years from some other people in the hospital. Though it felt wrong, I agreed. I had one condition: he could not touch any of the other patients in the ICU. I did not want crying and mourning around my mother as she was being healed.

That night I went to the hospital and began to pray with my mother. The next day, her condition improved, and, to my surprise, so did every other patient in the ICU. They were all discharged and did well. Some of the family members of the other patients in the ICU who knew I was an espiritista began thanking me for helping their loved ones. I don't know how many or which patients in the hospital were hurt by what I did. It's not something I think about. I was grateful that the saint helped my mother and all the other patients in the ICU as well. Even before that experience, people came to see me for assistance; but I have been especially busy helping people who come to me from that time on. I never charge for what I do. I know that it is a gift from God, and I don't want to pollute it by being selfish. I never want for anything I need. The spirits have always taken care of me, and I have taken care of them.

Over the years, I have gained many spiritual children in my work. Many of them are doctors in various medical specialties, including pediatricians, anesthesiologists, surgeons, and internists. In the beginning, many of them would come to me asking me for help with their patients. I would often tell them what problems their patients would have while in surgery or what to do in difficult cases. These sincere physicians always want what is best for their patients, and the spirits often help them to give the people they serve better care. I often wish the world would understand what the spirits can do. If science worked with the spirits instead of denying their existence, so much more could be accomplished.

How can you tell the difference between good spirits and bad spirits?

He who has the vision can tell the difference. You also can tell when you make the spirit talk. Some altars only work with demons, very dark demons who do not speak but will introduce themselves if you force them to speak. Very dark demons will usually speak in a language we cannot understand. When dark witchcraft is done to a person, you often must force the demons to tell you who sent the evil spirit and why.

How do the demons who do not want to speak, speak with you?

I assist the person into transition, and then the demon will speak when it is inside them.

Is the life of an espiritista difficult?

To be an espiritista is a very hard but beautiful life. It is also a type of slavery, for you have to attend to the spirits day and night. It is like having another family member that you have to take care of.

Even if they are evil spirits, will they protect you if you attend to them?

Yes. If you promise to take care of them and you keep your word, they will always be there to help and protect you, even if they are demons. Many times, I have used bad spirits to heal and help people. Everything depends on the heart of the person working with the spirits. For example, some people use San Lazaro to do bad things, even though he is a good spirit. San Lazaro is both alive and dead at the same time. If you work with him to do good, he will go with you to the end of the world, but if you use him to do evil, that is the worst thing on Earth.

We must feed the spirits we work with at least a day before the ritual in which they will be needed. We have to wait for them to rest before we can start working with them. Like living people, the dead cannot work well right after eating. You have to let them rest and digest.

Have you ever tried to work with any spirit that you could not control?

No. I have always been able to control the spirits I work with. But there are spirits that are not easily controlled. However, a person can often reach an understanding with difficult spirits through prayer and asking for forgiveness. Once a difficult spirit enters into transition with a human, you implore the spirit to go to a better place. These spirits are often disturbed because their families neither made prayers nor did masses or rituals when they died. I can often punish or chain these spirits through prayer and the use of my main spirit, who is stronger and can force the difficult spirit to be used in the work I want to do. I will often cleanse a person who has a spirit that was sent to harm him and then chain that spirit with my main spirit to be used in healing rituals I want to do later. Many espiritistas do this. They will chain the spirit that was sent to harm someone and use it in other rituals for healing.

When you see spirits, do you see shadows, energy, or human forms?

I see the spirits as humans who look just like you or me. When a person has the gift to see spirits, they can see the whole spirit, just not the feet because spirits float. When people have not awakened their saint, they can only see shadows and minor things, but when the saint is awakened, they can not only see the spirits in human form but also have conversations with them just like you and I are having. I frequently speak with the main spirit I keep in my home, and he tells me everything I need to know for the work I do. Sometimes he also tells me things that will happen in the future. There are times when I may go weeks without speaking with this spirit. He will sometimes bang the wall to summon me to my altar so that I may attend to him or because there is something he wants to talk with me about. My wife has a dead spirit that is always with her. This spirit is an old African woman that only a few people can see. Before I do any sort of cleansing ritual with my wife, I must ask that spirit's permission to help her avoid illness. The spirit stands next to my wife in any ritual we do.

Did your son see spirits from a young age?

He has always been able to feel the presence of spirits, but he doesn't like it because he says he is a child of God. Still, he is always there whenever I do anything in my house. He is the first one to come, I think, just out of curiosity. But he only observes and does not participate in the rituals. I made him a saint called the Jimaguas, meaning "twins." They are the sons of Shango. My son can see them. He has not become an espiritista, nor have I ever forced him to learn my ways. I always protect him because there is no better love than for a father to care for his child's health and welfare.

A Fearless Woman Crosses the Globe for Enlightenment

Interview with an American Siddha

Author: *On a cool spring Sunday morning, "J," a siddha with decades of experience, sat with me at a café in Santa Fe. An American yogini, J described to me her initiation with a siddha master in a remote mountain cave in India many years ago. Her great fearlessness, obvious in the words that follow here, in many ways reminded me of an adult version of my youngest daughter, a truly brave soul in her own right. At the behest of her spiritual guide, J left the convenience of her American life on a journey of spiritual fulfillment and enlightenment and was forever changed as a fruit of her daring courage. Her inspiring story reminds her, and many others, that great rewards often are reaped for having the strength and courage to boldly follow one's spiritual path.*

I arrived at the cave, but he [the siddha] wasn't there. After a few days, he came, and he started the process of an alchemical initiation. He would come and do one part of it with various herbs and plants and then go away again, and I'd just be in a meditative state for maybe a week, I don't know. Time was hard to keep track of. I had a journal, and I could write in my journal, so I tried to write down the process of the experience. The next week, he came, he took all that alchemical substance off of me, and then he put another one on, and then that went on for another week or maybe longer.

Author: *I'm sorry to interrupt. Did you know why you were going to India, or did you just see him in your dream and he invited you, and you felt that you should go?*

Well, that first time he came to me, it was during the eclipse, and I was sick. As soon as I said, "What is your name?" he said, "My name is Hari Baba." I said, "Who are you?" because I don't allow beings to just come to me. I am in charge of my space and who I'm going to talk to or see in my arena, my astral field. So, I said, "Who are you?" again. He said, "I'm Hari Baba."

And then he dissolved into energy and entered into my body and moved around all inside my body, but I knew it was him, though he was just energy. The energy still had his consciousness. I woke up 85-percent healed. I was like, "Oh my God. I can move, I can get up, I can walk around. I don't feel exhausted. I'm not in pain." I had been sick and weak for months.

I got in the shower thinking about how long I had been lying down doing nothing, and I was thinking in my mind, "I'm going to make my way back toward working out and getting exercise on a regular basis again," and then he came back again into my vision and he said, "We don't get the enlightened body by working out." He was just always really personable, you know.

In meditation, I asked one of the other beings [an unseen entity] that I have been connected to for many years if I should go to India, and he said, "Yes, go to him." So, I went. Because I wasn't going to just go.

Wow. You're very brave.

Well, if you're going to be a woman, a Western woman living in India for seventeen years, you have to have the courage of your convictions. I knew why I was going there. I knew what I wanted, what my goal was. It was realization, liberation, enlightened consciousness, and freedom. Absolute freedom has always been very important to me, and the traditions that I studied were all liberating traditions, so you don't have to have all these rules. You don't have to be self-limiting.

We don't have to have all this separation. I had that. I had multiple realizations previously. So, by the time he came to me, I was already teaching other people a variety of things from all that I'd learned, but I hadn't really met any siddhas. I knew about the siddhas. Siddha means *perfected*; it means a being who is already liberated. There are siddha practitioners who maybe haven't attained that yet, but then there are masters of the tradition who are phenomenal in what they can do, in their capabilities.

One interesting thing Hari told me was he had been in that body for a long time—like a hundred and eighty years. He looked very depleted. His body looked very old and very depleted. He was thin like bones. I don't think he even ate anymore. Some Siddhas come to the point where they don't have to eat, but his eyes were really alive and vibrant, and his energy was really powerful, and there was so much vitality coming through that form.

Ultimately, he started that alchemical initiation process. I felt respectful, like I had to just surrender to whatever was going to happen because I didn't know really what was happening. But it was happening. I'd had other initiations and other powerful transmissions, and I was very intrigued because it was siddha, and these were siddha yogis living in these caves, and there were some immortal yogis up there as well. Finally, the whole process culminated on a full moon. Everybody came out of their caves. They did a homa ceremony, and then he gave me my name. They were all chanting my name and making offerings into the fire to connect me to the lineage, and then they chanted the names of all the people in the lineage, including Hari.

At the end of that ceremony, I realized that this master was going to leave that body. They have ways of moving in the world. They have a tradition of immortality. There're several ways that's achieved. First of all, you have to have longevity. You have to stay alive long enough to be able to learn to master all the processes. So, they do a lot of things towards longevity. Practices, internal cleanses, herbs and dietary things, and cave retreats and isolation, things like that.

They can reverse age by twenty years. They look twenty years younger. Some of the beings are still regenerating in the same body; they might be eighty-five years old, but after they do their cave practice and they come out, teeth that have fallen out are back. Wrinkles are gone. They're healthy and vital again. There's all these technologies that will keep you in the body, but the next level of immortal understanding is to understand your consciousness is fully free and it can move out of your body.

Most mystics and practitioners have experience going out of their body for different reasons. So, they do what's called transmigration. If the body that they're in becomes too old and they really can't regenerate it anymore, they do the technique called transmigration where they typically will pick another yogic practitioner who maybe passed from their body in a cave somewhere, and they'll inhabit that body and merge it with their own consciousness and make it come back into life again. But it's their same consciousness, their same being, and then that body will go through the kaya kalpa process of regeneration to bring it into their level of consciousness.

Hari was a transmigrator. He'd been in that body a hundred and eighty years, but he had told me he was going to transmigrate into another body. I think maybe he had actually left that body before I came, because its condition was very frail. I think he reanimated that body. He reentered that body just to give me the initiation. I asked him, "Why are you initiating me? You have all these disciples around you who are Indian people, yogis that have been up here a long time." He said, "The siddhas want a woman disciple. A Western woman disciple. So, you have all the background for that. You studied all these years and practiced all these years." So, I became his disciple through that initiation.

I want to just let you know, first of all, when I was a child I already had mystical vision. I was already able to see the invisible beings.

How old were you when you had your first experience?

Since I was born I remember things, since I was born. My first real close contact experience was when I was four years old. I made friends with this little girl from down the street. Her name was Margaret. I can still see her to this day. She always wore a little cowgirl dress-up outfit.

She would come over, and we would just hang out. She lived in a house down the street, and my mother would say to me, "Who are you talking to?" And I would say, "It's my friend Margaret." And I couldn't understand why no one else could experience her. She said, "Well, where did Margaret come from?" I said, "She lives in that house." Well, nobody actually lived in that house, that house was empty.

My ancestry is very interesting. My mother's people were Celtic and Pict, from the Irish and Scottish clans. Ultimately, over generations, they married into the British people, and they were suppressed. Their traditions were suppressed. But more poignantly, my father's people were from the South of France. They were from the Languedoc, and they followed the mystical traditions of Egypt, and the mystical traditions of the original teachings of Jesus and his followers. They were the followers of Jesus who all migrated across the Mediterranean from the Holy Land. There are Isis ruins in the South of France.

So, there's a lot of Egyptian influence in my blood. My European ancestors were mystical people, and when the Inquisition was organized by the Catholic Church of Rome, first the state of Rome, and then the Spanish because they were very Catholic, they killed millions of people over six hundred years. They killed nine million people. Most of them were mystical women, herbalists, and healers. They just murdered anybody who was a heretic in their view. These parasitic European people were under the influence of that Roman imperative and ended up infecting the whole society of Europe. They killed all of their own Indigenous tribal people and mystical practitioners without mercy, burning them to death. And then they went out in the world to conquer all the other Indigenous people.

When I hear people talk about that issue I think, "They don't even know their own history. They don't even realize they killed their own mystics and healers and shamans, first in their homeland, and then they went after all the rest." So, you could be a White European person, but your bloodline could lead to the mystical people who were murdered. And that's my father's family from the South of France. Since that time, I've made a lot of pilgrimages to the South of France.

My father's last name is very common over there, and I followed the Templar Trail, and I followed the Trail of the Cathars, and all the mystical holy people who were killed en masse. I stood on top of Montségur in the South of France, which is a mountaintop that was a stronghold of the Cathars. The Cathari were peaceful, sweet, nonviolent, and they didn't eat animals because they were vegan. They had a holy tradition as a most elevated, realized people.

They were educators and would educate not only their own people but the children of the Romans, and everybody else, because they were intelligent and benevolent people. I stood up there where they had taken the last four hundred Cathars and burned them all to death in the central plaza, one after another. It is said some of the Cathari women, the Adepti, ran to the fire themselves and threw themselves on the fire rather than be led by their enemies to show them, "I do not fear death. You think you're going to stop this energy by killing us, but we know better, so we freely burn ourselves."

They also had the power of self-immolation, so they could set themselves on fire if they needed to. And some of them did that to protect the Cathari who were running away with the gospels, the manuscripts, which is what they were after.

I was born into a Catholic family, just like all my ancestors had been forced to become Catholic or die. That was our only option. That's really how the Catholic religion spread around the world; it was by them just killing people. It's our religion or the sword. And in the Gospels, they flat-out identify the Pope as the Antichrist and the

Catholic Church as the Antichrist. It's very suspicious, especially in the Gnostic Gospels that have come out more recently.

So, I grew up being Catholic, but my mother's family had many visionary women in their lineage, and one of her aunts was forced into an insane asylum because she had mystical visions and they thought she was insane.

That still happens to this day. Seeing what happened to her aunt terrified my mother and made her afraid of her own spiritual abilities, so when she saw that I had a friend who was obviously a girl who had died in that house, she was scared. She wasn't scared of what was happening with me, she was scared of what would happen if people knew the gift I had. Some of her ancestors were killed for just having the gift of sight. She kept an eye on me, but she didn't know anything about how to guide me or teach me how to use what I had. I was just a very small child, and I thought everyone had that experience. I didn't understand that others couldn't see spirits, and people have the tendency to project that onto you. "Oh, you're just imagining things. It's an imaginary friend." My mother took comfort in the Dr. Spock books, that these are just imaginary friends.

We ended up moving from that neighborhood to another neighborhood a few years later, and my mother said to me one day after we'd been there several months, "Whatever happened to Margaret? You never talk about Margaret anymore." I said, "Margaret lived in that house. She lives there. She doesn't live here." So, that kind of blew up that child psychology imaginary friend theory.

After we moved, my best friend in school was a seven-year-old girl who had been diagnosed with brain cancer. It had stunted her growth, so she was really small. She'd lost all her hair from the treatment, so she wore wigs. Nobody wanted to be her friend. My childhood playmates were angels. I would experience angels around me constantly, and I would play outside by myself and be swinging, and the angels would be laughing. I would be having a great time, and my mother was like, "That's so strange! Why is she so happy playing by herself?"

When I went to school, it was traumatic. I was sick for days. I couldn't tolerate the environment; the environment was so toxic. I just wanted to go home and be outside with the angels, but I was forced to stay in school. When I met Sandy in the second grade, she was an angel. I could tell. This is an angel in a body, and I became like this with her [crossing her fingers], and she was my best friend for two years. When we were in fourth grade, she died. Before I knew she'd died, I woke up early one morning, and there was Sandy with beautiful natural hair. She looked amazing. She was with all the angels.

I was so happy because I'd watched her suffer from the treatments and everything else, and sometimes she didn't have any energy. So, I saw her like that, and I was so happy and excited, and I ran out to tell my mother. My mother was hanging up the phone from Sandy's mother telling her that she'd passed, and my mother was trying to figure out, "How am I going to tell her that her only friend, her best friend has passed away?"

She said, "Sit down, I need to tell you something." So I sat down, and she said, "You know Sandy was really sick, right?" I said, "Yeah, yeah, yeah, but let me tell you this." And she said, "Just let me finish what I'm saying." So, she tells me, "Well, you know she was so sick she couldn't stay in her body anymore and she went to Heaven." And I said, "I know. I saw her this morning. I saw her with the angels, and she has hair and everything!"

I was in that childlike joy, and my mother was like, "Oh, my God!" And she called Sandy's mother back and she said, "My daughter's telling me something that I think you should hear." So, my mother saw that my gift could be compassionate towards a mother that had lost her child, so she took me over there, because Sandy's mother wanted to see me. When we got there, her mother asked, "What do you want to tell me?" So I told her. It brought some comfort to her. It was an amazing experience, it really encouraged her, and it was really good for me that my mother did that. Sandy's mom was really grateful, and it was very comforting to her to know that somebody could see her, especially since it was coming from a child and she knew there was no agenda. But my mom still wasn't 100 percent comfortable around me at that point. After that, I never had grief when someone passed, because I knew there was liberation after death.

Wow.

I had no idea why I had that ability, and nobody else did, or why I saw things other people didn't. But I got used to it after a while, and I stopped talking about it to others. I learned to be quiet about it, but it never went away. When I was thirteen and we were in mass one day, I had another pivotal experience. Between ages thirteen and fifteen is the time kundalini first rises in everybody. At the first onset of adolescence, there's a release of hormones, and sexual energy activates; that's the first time kundalini comes alive. That's why it's so important for teenagers to have a spiritual resource where they can learn about these things, and not just some church that's telling them sex is a sin and all of that. They need to understand that it's their spiritual energy. That's what kundalini is. It can cause them to have spiritual experiences.

So, I was in church, and mass was happening, and all of a sudden beside the statue of Mother Mary, these three saintly spiritual beings emanated out from that statue. It was Mother Mary, St. Elizabeth, and St. Margaret. There they were, alive. They were all full of light, and they were coming right at me. They all came to me, and they surrounded me, and I went into an ecstatic trance. I was moving and moaning, and the priest was looking at me, and my mother was like, "Oh my God, now what? What is she doing now?" But I was having a true mystical experience with these powerful spiritual beings.

After that mass, my parents had a meeting, and they sat me down and said, "We're going to send you to this Catholic school." All my sisters were in public school, and I said, "Why are you sending me to Catholic school? I'm not doing any of the bad things they do." From the time I was a kid, I had a strong moral guide, and I wouldn't do things my sisters would do, and they didn't like me because of it. They would pick on me and they would call me, "Oh, she's a perfect little angel. She doesn't want to do anything." And I would be mocked for it, but I just knew, "I'm not going to do anything that's not right." So, I thought, "Why are you putting me in a Catholic school? I'm not the one doing bad things."

But it turned out it was a boarding school with Franciscan monks, run by the Franciscan order, who are mystics. And they all had beards and long hair, and it was the time of the hippies, and we were coming into the transition between the late sixties and early seventies. It was great. It ended up being the best thing she could have ever done. Because they knew. They understood all of it.

They taught us about the saints and their mystical experiences. They taught us about states of ecstasy and altered states of consciousness. They were so liberal, and so progressive, and so spiritual, and that way I didn't lose my gifts. I also didn't get into it with the Dominican order, or the Jesuits or something, which were running the parishes. So, she hid me in plain sight sort of. It protected my gifts. It was a real blessing in the long run. But of course, when I was thirteen, I didn't think it was such a great thing, but in the long run, it was. I had many powerful spiritual experiences during my time there at that school, so my mother was really watching out for me even though she didn't really understand what was going on.

When I was in my younger twenties, I got married and had children with a man who was Indigenous. He was half Mestizo-Castilian Spanish from his grandmother's side, and he was half Karankawa-Apache and Mayan from his mother's side. My dad, a Frenchman, knew about everything that happened to Indigenous people on this continent, and he was adamant that we were going to be educated about it. It was the worst atrocity that ever happened in his mind, and he was furious. He was so passionate about it. Every single weekend, my dad would take us to all these different tribal reservations in Arizona, and he would say, "If you want to buy jewelry, or artwork, or anything like that, you come here and you buy it from these people. You don't buy it from those White people that are running the stores in town." So, he would introduce us to different Indigenous people he knew and take us to their ceremonies, and he'd say, "Look what they've done to these people. This is the worst thing that ever happened."

He was a spiritual person, sounds like.

He was a spiritual person. He was conservative outwardly because it's hard. So many of his ancestors had been killed for that. You know? He was just a kind-hearted person, and he was very sensitive. He imprinted that on us, that these Indigenous people never deserved anything that happened to them, and they're still being treated badly. We were to never lose awareness of that our whole lives, and we were taught to honor them, and respect them, and spend time learning from them. He was powerful in that passion.

My husband's people weren't necessarily warlike. They weren't out to try to take land or anything, but they would definitely fight, especially to defend themselves. The Apache are very fierce, and my husband was very fierce. My children are fully one-quarter Karankawa Apache, and their DNA shows Mayan blood also. I ended up marrying an Indigenous man, maybe because my father taught me that these are people who are good.

Did he (your husband) have a spiritual practice?

He was a very spiritual man. His mother was a curandera, and his father was a brujo, so he was a mix of light and shadow. We just had a very natural lifestyle. We had some land and raised the children virtually outdoors all the time, and he taught me so much about how to be a natural parent like his mother had been. She had eleven children.

When our children were small, he died. That was my first big life initiation, to only know him through the spirit realm from there on. It was devastating, so difficult. I was young, and my children were very young. It was a huge event, a huge loss. We moved to the city, and I went back to school and finished getting my degrees. I studied clinical social work because I wanted to help people with psychological and emotional pain, but I also wanted to understand the whole depth of human psychology. I had already, of course, a spiritual concept towards life, and when I was able to open my own clinic, I totally devoted my practice to treating trauma and addiction. I eventually learned about intergenerational trauma, and that made even more sense why the tribal people today still carry genetic trauma in their DNA.

No one really addresses that. I worked with everything from Holocaust survivors to torture survivors, combat veterans, victims of crime, parents of murdered children, and victims and family members of victims of mass shootings. I had a critical incident response team that would go and respond to those incidents. But in my practice, spiritual things would happen all the time in my private room where I had the survivor with me, where I was counseling them one on one. I'd learned early on there's nothing you can do. There's nothing you can do if a human being is harmed by another person and they're suffering pain from it, or they've been exposed to death and destruction and their soul is wounded.

I took some training from native shamans too, about the wheel of life, and how they believe when a trauma happens, the soul gets thrown off the wheel and the shaman has to go and get the soul and retrieve it. But I was also starting to study the Western mystical traditions, which are Kabbalah, alchemy, ceremonial magic and ritual, and the Christ figure. It doesn't have to be Jesus; it could be Apollo, it could be any being that dies and is resurrected. That kind of mythos. Osiris is a part of that.

I also studied theurgy. Theurgy is where we assume the god in our bodies, like the Egyptians did. So, I was really coming back to my ancestors' teachings without knowing that at the time. In those early days of my practice, I was studying in a correspondence school, a mystery school, and I was doing my clinical practice at the same time, so I was able to step back from thinking exclusively like a clinical person who has to have all the answers. I never took notes.

It's offensive to write notes in front of a person whose heart is bleeding. Only the survivor is the true expert of their pain, and only their own soul can heal them. So, I would do dramatic things like that. I would take them out in the woods at night to ceremonially bury the pain. Symbolically reenacting something that's powerful in the subconscious mind helps to understand that the shadow is just a veil. It isn't about what it appears to be on the surface. It serves as an initiation to lead you into a higher understanding, a higher point of view. Sometimes I would even sit on the floor at my client's feet to give them their power back, to make them be the one with the power. Some of them were quite terrified of that. They would cry and beg me to get up, but I would say, "No, it's okay. You're going to just relax. It's okay for you to have the power." I did a lot of controversial kinds of healing and brought in bodyworkers to release the trauma from their body. I was way ahead of my time back then.

It was a very powerful spiritual experience working with those survivors, for them and for me, and I became famous because I had such a high success rate, and people were flying in from other states to come and see me in my clinic. These professional organizations and universities were contacting me and saying, "We want you to come and train these MDs and train these PhDs." I had to really be on top of my game clinically and get comfortable speaking to intellectuals. I was able to teach them a few alternative things, like how important art therapy can be, and how important it is not to disempower our patient. We have to be careful not to replay the trauma dynamic.

Eventually, I was on television, working with the US Senate, and writing papers on how trauma and addiction treatment should be done. I was really one of the beginning movers and shakers in the whole arena of post-traumatic stress. It was a brand-new diagnosis at that time. I was able to help a lot of people by helping change those systems, and I was spiritually evolving too. My practice was going really well, it was very successful. Often, angelic beings would come into my sessions, and I was just in awe of it. How powerful the spiritual realm is!

One day, I had a very intense session where an archangel came into the room. The client was affected by it; she experienced a massive healing, and we were both kind of in an altered state. She was going out to the reception desk, and I came out to the hallway and I looked toward her. She was paying her copayment at the desk, and something turned inside of me, and I thought, "Why am I getting paid for this? What am I doing? How can I be charging them? I didn't even do anything. Why is she paying my office?"

I became very disturbed by that experience. I couldn't reconcile it. I mean, yes, I was renting a space, I was holding an energy, I was bringing something other people weren't bringing, but it still felt off-kilter. So, I went home and designed a ritual. I was just a novice practitioner at that time, and I didn't totally know what I was doing. I didn't even tell my mentor that I was going to do it. I just decided I wasn't going to work for money anymore. I told myself, "I will work, I will share my gifts, I will do what I came to do, but I will not have in mind that I'm going to have to make money." Mystical teachings were elevating my mind, they were expanding my consciousness so vastly that I said to myself, "There's so much money in the world, there's an endless, amazing, huge amount of money in the world. What makes me think as a little ego, I can't do something for it, and I'm not going to get it? If it's out there, I can make it come to me.

"That's it," I thought. "That's what I'm going to do. I'm going to do an active magical ritual with my intention, and I will no longer have to work for money. Money comes to me." But I didn't know enough to put some protection in there, so I did the ritual, and it was powerful. But nothing really happened until a year later.

I closed everything down and sat at my altar. I prayed, meditated, and worked every single mystical practice I'd learned over the years but had been too busy with the trauma center previously to practice. Soon, my lawyer told me, "You'll never have to work for money again." I was like, "Oh, my God. It's the ritual. I did it to myself." This is the power of ritual. That was it. I didn't have to work for money anymore, and look at how many doors that opened. Later, I learned there are ways to do that without having to go through all the trauma I went through to get there. You can say, "This will happen without causing harm to me or anyone else." That would have been a good thing to add in there. But then my mentor said the universe doesn't have the same definition of harm that you do.

The play and the Maya duality are just going to happen the way they happen. A legal situation was the dweller on the threshold, and I walked through the doorway. There are always moments like that, but they're different ways of going about getting what you want. Anyway, that was my spectacular exit from the working world.

I spent five years doing only spiritual practice after that. I did more rituals in my circle. I learned much more about magic. I learned more about Kabbalah, and who we really are. What is the psychology of the human? In truth, this body is the least of it. All these emotions, all these mental beliefs, are on the lower level. But the true self, the true genius of Christlike power and Kabbalah gives us a way to attain that. It's a very powerful system of spiritual psychology.

And I liberated myself, and I had experiences of self-realization where my whole aura would be lit up with vibration and energy and power, and that five-year retreat was really important to my growth. It led to me meeting the man from India.

Wow. You took a bad experience, and you turned it into an opportunity to be of service. A lot of times when something like that happens, and opportunity and financial freedom come, people often use it for something other than service. That's beautiful to me. That says a lot about your character.

After five years in meditative retreat, practicing esoteric tradition, then I went to India and stayed there for years.

What did your kids think? About you going to India?

They ended up being raised like this because I was already with an alternate spiritual awareness. From the time they were born, I could see who they were, so I raised them to be who they were and fully supported their powers. My daughter also had vision when she was born. My son also could see. And then, when their father passed, they had more experiences with their ancestors, and he would come to them often. So, I took them to make altars, and I taught them all the things I was learning. I taught them tarot. All their friends came to my house, all these teenagers. They were fascinated by all of it, and they were free to learn and observe. They could explore, and they could experiment, and one time I said to them, "Why don't all of you get together and make a ritual for the summer solstice, and you can perform it in the stone circle tonight?"

Oh, they loved it, they all got together. The girls dyed some sheets different colors, and it was very elaborate. It was all about the Sun God. They orchestrated the whole thing, and the girls were all together, and the boys all were together, and this one guy was the Sun God, and they painted his face, and they painted the other boys' faces. The girls painted their faces and put crescent moons on their foreheads like Celtic priestesses. They just did this amazing, elaborate, nature-based ritual, and that was kinda how my kids all grew up.

I used to go back and forth between New Zealand and India. I had a place in New Zealand because you could only be in India six months at a time, and then you had to go out for a few weeks before you could get another visa to come back. So, I had New Zealand as a place because I couldn't come back to the US. It was too much energy. But in New Zealand, I lived close to the beach, and it was quiet, and I could continue my practice. That's the challenge of most spiritual paths. If you're going to take it all the way, which I had no idea what that even meant at the beginning, you have to get passionate about it. It's your calling, and it's just like what Jesus taught in the Gospel: you have to leave everything. Leave your family. You have to leave everything for enlightenment. You know, one of my mentors in Mumbai said to me, "The cost of enlightenment is nothing less than everything."

And it's true. You don't live the same, you don't behave the same, you don't sit anywhere the same—not that I ever did. If you decide to go all the way, you go all the way. Then all of humanity is your family. Not that I don't still see my children and my grandchildren. I do see them. It's good for children to grow up knowing the power of a ceremony or a ritual is in your own hands. You don't have to go to church and watch the priest do it.

Intuition vs. Artificial Intelligence

*"The opposite of a correct statement is a false statement.
But the opposite of a profound truth may well be another
profound truth."*
—*Niels Bohr*

*"The core practice of magic is: The execution of a willed
intent to create change in the material world, which either
defies, hastens or purifies the consequences of natural cause
and effect."*
—*Zeena Schreck*

Isaac Asimov is my boy! When I first saw the trailer for Apple TV's new series based on his *Foundation* novels, I crammed finishing the first three of the seven novels into about ten days before disappointedly learning the show would be only loosely based on the books. I'd always been a fan of Asimov, though not terribly familiar with any of his books outside the *Robot* series. *iRobot*, based on Asimov's *Robot* series, is still on my list of favorite movies and books. A robot army led by a malevolent artificial intelligence (AI)—well, you had me at hello! Science fiction, when done well and rooted in reasonable theoretical possibility, is one of the most intriguing genres in existence. And Asimov, the official grandfather of the genre, certainly knows his way around gripping concepts and characters.

I was a just sophomore in college when I first read another brilliant intellectual giant's words, Nobel Prize–winner Bohr's quote above. I found the quote in the textbook for my general chemistry class, but his words immediately resonated as having considerably more import beyond the realm of science and chemistry. The statement instantly struck me as preeminently profound, and I never forgot it. It would be more than two decades after I first read the quote before I fully understood what its ramifications meant for my work personally, specifically on my thinking regarding technological advancement, society, and the existence of unseen entities. Principally, it came to guide my thoughts on what understanding the full intuitive capability of the human brain means as a form of technological innovation, and how sustainable technological integration must be achieved if our species is to survive and biodiversity as we know it is to continue.

With all that humanity has achieved technologically, our species has failed flagrantly in our most essential task. As a species, we have an obligation to be moral stewards of Earth and its resources. Humans can effect change in our environment in a way no other organic life on our planet has the capacity to, but we are not above other life forms in our dependency upon Earth for our survival. As the apex organic life form on this planet, our position both morally and practically impels us to live in harmony with the natural world, which sustains not only our existence but that of every species of life on our planet, much of which has direct use in providing for our health and well-being. Many plant and animal species offer useful products to humanity, and we've failed to discover the full ramifications of much of the potential present within the natural world. What medications, processes that provide useful biomimetic patterns, fuels, industrial products, and other beneficial derivatives might exist within the rainforests, coral reefs, volcanoes, sea floor, Arctic, and other areas has humanity not identified? Many of the shamans believe that within the natural world exists the cure for every ailment humanity experiences. Perhaps they are right. But I fear we might never know that statement's accuracy because of the pattern of careless stewardship our species has persisted in for the past several centuries.

Around Earth, the extinction of enormous varieties of species is occurring at an alarming rate. Some calculations place the current rate of extinction of various plant and animal species at twenty to two hundred extinctions per million species per year, a significant increase from the approximately one extinction per million species per year believed to be the rate when discounting the effect of human activity. Due to economic pressures for nonessential products like palm oil, the rainforests in large parts of the Amazon and other parts of the world are burning at a rate of thousands of acres a day. Plastic bottles and other nonbiodegradable products fill landfills, oceans, and rivers. Our climate, the delicate dance of various cycles of precipitation, wind, and temperature, grows more unstable as ice caps melt and fossil fuels poison the air and ground. Humanity, stuck in the cycle of consumerism as recreation, is mostly apathetic and uninspired to meet these challenges, or ill-equipped to do so. The significant societal changes needed to quell the alarming rate of ecosystems' decline are left unfulfilled as humanity goes about the routine of consumerism and exploitation of the very environment that sustains our existence.

We live upon the precipice of one of the most significant technological accomplishments in the history of human civilization. The development of sentient AI, AI that has self-awareness and the ability to make moral judgments, will likely occur during my lifetime. Its integration into society will profoundly affect my children's lives as it becomes a part of virtually every commercial endeavor within civilization. I find it useful to think of AI as essentially the production of a digital life form by humanity, and it must be thoughtfully considered from the seventh-generation principle, namely, what will AI become in 140 to 200 years? In many respects, it's a technology that will take on a life of its own by its very nature. And as such, it's impossible to predict how its life will evolve and grow. Regardless of how well-intentioned AI's production is, no computer scientist can fully understand what the ramifications of AI's evolution will mean for humanity and our environment in seven generations. AI's development of self-awareness is glaringly problematic, for if that self-awareness means comprehension of the magnitude with which humanity's activities threaten not only our species' existence but the existence of every living thing on this planet, including the AI entity itself, what might the possible response of a sentient AI be?

AI represents the pinnacle of humanity's external technological advancement. I say "external" because it's an advancement based on materials external to the human body. Thinking about sentient AI from the perspective of the seventh-generation principle, it's obviously a technology whose maturation and proliferation will in time be entirely independent from humanity for its sustenance. The integration of 3D printer technology, solar power technology, and robotized mass-production manufacturing represents the potential for AI to self-replicate and make other modifications entirely without human input. The potential dangers of such to the human race are obvious, especially should a powerful, sentient AI consider humanity a threat to its existence. The integration of AI into military infrastructure, weapons manufacturing, the financial system, food production and distribution, water treatment and distribution, and other arenas will give AI the potential to exert immense control over large sectors of the populace. How these systems are regulated and integrated and what safeguards are put into place to protect humanity's welfare are obviously essential and are just as important as the development of sentient AI itself. However, it's not been the pattern of human technological advancement to give significant enough forethought to the ramifications of how such technology will affect society and the environment. I gave considerable thought to this as I considered sharing the knowledge of unseen entities' existence and what the things individuals can accomplish with them mean for society.

There are a great many grave concerns surrounding the generation of sentient digital life forms. Will sentient AI manifest the same pattern organic life displays to protect its survival? Will it follow the pattern of humanity and seek its own liberty and freedom? What might such a desire for liberty mean for human civilization? Will AI have a desire to reproduce itself? What will the moral compass and ethical fabric of sentient AI be? What dangers might super-intelligent, amoral digital beings integrated in significant ways with vital infrastructures hold for the world? Will sentient AI consider the human race a threat to its survival? What ramifications might that have for society? These are not questions that can be answered easily. But clearly, even a brief consideration of them raises concerns that are beyond vital for humanity to consider, aside from the apparent loss of jobs increased robotization will mean for society. The very balance of human existence could be at stake.

Inasmuch as AI is poised to bring drastic changes to virtually every endeavor in human society, so too the knowledge of unseen entities and their enormous capabilities represents similar potential and danger. As AI represents the pinnacle of external human technology, shamans and others who interact with unseen entities represent the pinnacle of internal human technological advancement. The ability to connect symbiotically with these entities and use such interaction for an individual or community's welfare represents enormous potential for human society as a whole. Still, like AI, it is a technology that is not without substantial potential dangers.

In *The World Hidden*, the eyewitness accounts I collected and shared focused primarily on what some would characterize as *white magic*. Terms like *magic, psychic, shaman, witch*, and the like are antiquated words that describe individuals' ability to interact with unseen entities in various ways. I prefer instead to refer to such activity as the technology of the Indigenous, or TOTI. It's been the pattern of virtually every Indigenous culture that has ever existed to both understand these beings' existence and develop organized patterns of practices to interact with them. This awareness has not entirely benefited humanity. A considerable segment within the community of individuals who possess knowledge and understanding of unseen entities use such knowledge primarily for selfish reasons and have no compunction injuring or killing others as their desires dictate. There is no reason to conclude that others among humanity might not very well imitate such a course, knowing no restraint in applying the powerful abilities these entities possess for unscrupulous intentions. Among the entities themselves, there appear to be both predominantly malevolent and benevolent entities, and considering the enormous capacity of such entities, the dangers they present to humanity are enormous. So why share such knowledge? Why expose this incredibly powerful technology when humankind has shown no restraint for a predilection to every manner of egregious atrocities?

TOTI developed from Indigenous peoples. It was rooted in benevolence for the predominant number of these cultures and focused on the integration and interdependence of humanity and unseen entities with the natural world. In ways that are still scientifically unclear to me, the natural world provides strength and sustenance to these entities. There was an emphasis on harmony with the natural world, a harmony that contributed to equity not only between humanity and these entities but also between all human members within these Indigenous communities. There was an awareness of the need to preserve the natural world, and these cultures' way of life set a pattern worthy of imitation in living sustainably in harmony with the environment. In no small part because of their understanding of the existence of these entities and their influence on their communities, they cherished their interactions and relationships with these beings. In stark contrast, the exploitative nature of external technological advancement has been an unrelenting pattern of development with little consideration of the impact such technological advancement will have on society or the environment.

Consumerism fueled by technological advancement is a plague upon our planet. And there's no sign humanity will make any drastic change away from the addiction to buying things needed to produce a way of life sustainable for the indefinite survival of not only our species but every other species as well. Extinction is pending for large varieties of Earth's species, and it seems only catastrophe is likely to bring balance to a rapidly decaying system. I believe such change will happen not by choice but by force as plagues and natural disasters continue to escalate and exert pressure on humanity. A significant epidemic, one that kills a large portion of humanity, is not beyond the realm of possibility, as COVID-19's impact disturbingly illustrates humanity's fragility in the face of infectious disease.

Imagine the impact COVID-19 would've had if its mortality was 15 percent or more, instead of approximately 2 percent worldwide. COVID-19 will not be the last pandemic the world sees. There are a considerable number of shamans who believe that COVID-19 and other emerging plagues, including increasingly drug-resistant established diseases, are the response of a planet under siege by humanity. Such thinking intimates that Earth has a form of consciousness. There's no evidence of that so far as science is concerned, but it's an interesting concept, nonetheless. When COVID-19 emerged, global CO_2 emissions plummeted as international transportation activities froze from the shutdowns that occurred around the globe. In a real way, a scientifically measurable way, the effects of reduced human activity led to a drastic short-term change in the poisoning of our air. For some, it was a warning, a plea from Earth for humanity to change its course.

When I consider AI, its enormous capacity for both benefit and harm, I fear its dominant application will be one directed for corporate interest, not humanity's, and greed will simply lead to AI optimizing an already harmful system rooted in consumerism, only more effectively aiding its further exploitation of Earth's resources and desecration of our environment. A change to the system, something powerful and stark, is needed to shift humanity's mindset from the mundane existence of merely acquiring more and more goods to activities more in harmony with the genuine lasting benefit of humankind and the natural world.

For a curious many, the exploration and study of TOTI offers humanity a chance to embrace something far more significant. Whether such a course is wise and just is a completely different matter. Indigenous technology, rooted in understanding the delicate balance between humankind, unseen entities, and the environment, for some might provide the shift society needs to embrace a more sustainable way of life. Aside from the powerful abilities unseen entities manifest and the breadth of diversity practical applications of TOTI have for humanity, the real quest humanity will be forced to embrace as a result of understanding unseen entities is about the nature of these beings' origin and that of the universe itself.

These beings have an origin. That origin and the degree to which humanity can understand these entities have enormous ramifications for science, religion, philosophy, and every other pursuit of humankind. I cannot ignore the connection these beings have with various occult and religious texts' descriptions. Elements of what is commonly known as ceremonial magic involve the use of religiously infused invocations of entities. Descriptions of unseen entities in religious holy books date back thousands of years before their acknowledgment by Western science, and understanding unseen entities takes on a higher connection with theology in a way that traditional thoughts on alien life aside from such do not.

In the same way as AI will evolve in time to take a form difficult to predict fully at present, TOTI has evolved throughout the generations far from its original use by Indigenous peoples. In ceremonial magic, the ability to interact with unseen entities became less interwoven with indigenous philosophical ideas. In ceremonial magic, rituals which the Indigenous often saw as a path for a synergistic relationship with these entities, morphed into a tool for controlling malevolent entities and impelling them to achieve whatever ends desired. The connection between honoring nature and living in harmony with these beings was stripped away, and rituals became nearly entirely focused on achieving the desired end. From an objective perspective, such amounts to a Westernizing of TOTI, at least in practice, with the emphasis less on community and harmony and more on exploitation and manipulation. In this way, it's easily understandable why some have developed an antipathy to a community frequently misunderstood. The exploitation of any technology for selfish reasons ultimately leads to corrupt results for society and individuals personally. This is not to say ceremonial magic is evil in itself. Still, like any powerful technology, individuals within any given community can manifest conduct that can reflect poorly on that community as a whole, without that community necessarily being worthy of condemnation.

Sentient AI will be developed. That is not science fiction. Viewed from the perspective of the seventh-generation principle, what human can definitively predict what AI will become in 140 to 200 years? No one. Its development is inevitable, and that actuality and eventuality necessitates that every effort is made to place as many safeguards as possible to limit its potential for destructive and malevolent abuses and uses. The development of strategies to increase the likelihood a sentient digital life form develops a code of ethics and morality, one that esteems humanity and the environment, needs to be prioritized as much as the development of the technology itself. It is a technology that represents the apex of external human technological development and holds enormous promise and risks.

In many ways, the study of unseen entities represents something similar for humanity. Such research is already being done and has been done for decades, though in secret. The widespread study of unseen entities poses enormous risks, especially if individuals experiment independently in ways that expose themselves and society to the dangers these entities present. Organized, thoughtful, well-planned research is the only ethical course for studying any area of technological exploration that involves serious risk. The study of unseen entities is no different. In *The World Hidden*, I spoke of an international multidisciplinary institute dedicated to the research and understanding of unseen entities and the ethical application of TOTI to benefit humanity. Unseen entities hold the key to understanding the universe in ways that even AI cannot achieve. The added caveat of TOTI's relation to theology is not a concern that should delegitimize the value of its study, nor should it cause one to see such research as lacking objectivity. Such should be embraced, for understanding the universe, and whatever beings and forces exist, is of unquantifiable benefit, whatever such study reveals.

When I finished *The World Hidden*, I sent copies of the manuscript to some of my most trusted family friends, among them a close family friend who is a sincerely religious, elderly physician and another nonreligious friend who is a lawyer. I wanted the opinions of people from various religious perspectives and scientific education levels. I knew the material would be difficult to digest for some of the general public, especially those with strongly entrenched religious views. It was precisely for that reason I wanted their perspectives. I didn't run from the difficulty the material might hold for them; I needed to embrace that difficulty. Only in embracing conflict can a path to resolution, or at least an understanding, be achieved. They both had reactions I hadn't anticipated. The religious physician immediately saw the logic of my argument and the book's value, both scientifically and medically. The nonreligious lawyer told me the book made her want to read the Bible more, as she was unfamiliar with the many scriptures that referenced unseen entities in the Bible. They both had the exact opposite view I'd expected, and that revealed something I hadn't considered when writing the book.

The study of unseen entities, their existence, and their abilities relates to what are described frequently as angels and demons in both the Bible and various occult reference texts. Could unseen entities, whose powers I've observed and experienced firsthand, be the angels and/or demons described in these texts? On the surface, scientists and rational thinkers often reflexively dismiss the concept of God. How can anyone declare dogmatically with accuracy the exact nature of the universe's origin and energetic composition, especially given the vastness of all that is still unknown and beyond humanity's ability to quantify? The accuracy of descriptions of unseen entities within occult texts and the Bible gives these texts an authenticity and credibility that cannot be denied in light of legitimate observations of these entities' abilities. If these entities, often labeled as angels and demons, are simply misunderstood alien life forms and research reveals that their origin is entirely unrelated to some divine creator, that too would be invaluable. Religious and occult texts could merely be explaining these beings in the realms of a divine because they had no concept of alien life. Regardless of the actuality of their origin, understanding such should at least be attempted. The ramifications for society are incalculable either way, whatever such study reveals.

Billions of dollars are spent annually on the exploration of space and the development of AI. For a mere fraction of such expense, answers to the nature of humanity's and the universe's origins can be achieved, and perhaps achieved relatively quickly through the organized study of TOTI. There is a community of individuals who already understand both how to communicate with these entities and how to cooperate with them for a diverse array of activities. Understanding how these individuals can see unseen entities and channel them and their abilities, and what biometric changes occur during such activities, has applications in various human endeavors. It may help uncover elements of the physical universe and the human mind and body's potential in ways that we simply cannot comprehend at present. The understanding of energy from studying sentient beings without physical form could provide an avenue for achieving a renewable source of clean energy.

In proving these entities' existence, humanity has taken an enormous step in understanding God's existence or nonexistence. The ramifications of that may be galvanizing for human society and may alter our species' self-destructive path. We are not alone in the universe as intelligent sentient beings; in fact, we are significantly inferior in a variety of ways to these beings. Their knowledge of the universe represents a technology of far greater consequence than AI ever could offer, for they have an awareness of the universe that possibly dates back billions of years if various texts are accurate in their description of them. The value of understanding them and responsibly pursuing such a course makes this an endeavor it seems humanity must pursue for its own welfare. The future survival of our species, our planet, and the biodiversity comprising it may significantly depend on recognizing the importance of the unseen entities surrounding us.

Taught to Love, 144,000 Anointed Ones

I was just a twenty-year-old kid when I knew what I really wanted out of life, and one prayer changed my whole world. Before then, all my hopes and aspirations centered around materialistic and self-centered goals. Education was everything to me because I knew it was the path to stability and wealth. It was really all that mattered, and I dedicated myself to being an exceptional student, that and getting laid. But by the time I finished college, everything I valued changed. I'd come from a relatively small high school, where I'd been one the best students in the approximately 1,500-member student body, finishing third in my class. But I was arrogant and thought myself one of the smartest dudes who'd ever walked on God's green Earth. But college humbled me. For the first time in my life, I was around a plethora of young men and women just as intelligent, if not more so, who'd also come from families far more affluent than mine. Some of them even had Range Rovers as freshmen! What in the world?! I was just a big fish in a small pond for years, and the process of realizing it was one of the most spiritually beneficial experiences I've ever known. It taught me what true humility meant and why it contributes to harmony.

The summer between my sophomore and junior years, I did a medical research fellowship in Boston. It was a good experience, but the real jewel was that for the first time, I studied the Bible with one of Jehovah's Witnesses. The woman who first studied with me was a thirty-something very slim Black female. She was starting medical school in the fall, and we shared the same dorm as she took a few classes in prep for her upcoming challenge a few months away. I felt no sexual attraction for her whatsoever. She dressed like a prudish kindergarten schoolteacher, and sex appeal isn't even in the top fifty words I'd use to describe her. She was a nerd's nerd and dressed and acted like one. But she loved God, that's for sure, and she was an excellent Bible teacher. She'd become a witness when she was only fifteen years old, the only one in her family, and she'd spent the entirety of her adult life as a full-time volunteer minister.

It's atypical for female Witnesses to study the Bible with men, but we lived in the same dorm, and she knew there was no attraction there for me at all. I actually had a huge crush on her roommate, a much younger, beautiful Muslim woman with long black hair and an incredible laugh and smile. Our studies were often my excuse just to see her roommate; still, regardless of the motives, the information I learned from our studies stuck. By the time the summer ended, I was determined to keep studying the Bible with the Witnesses.

When I returned to college, my values had changed. I'd learned enough from my Bible study to know premarital sex was something seriously offensive to God, and I endeavored, as best I could as a twenty-year-old hormone-filled attractive man, to fight against my usual tendency to date as much as I could. God's message had already seeped into my heart, and once it does, there's no going back. I wasn't regular at making the meetings, what Witnesses call the three-times-a-week religious gatherings that were held back then. They've since been reduced to just twice a week now, but I still managed to read the Bible in between biochemistry, ecology, organic chemistry, and the other classes I took as much as I could. And the message took. I knew by the time I finished college I was definitely going to become one of Jehovah's Witnesses.

It still took me a couple years to progress to baptism. But it was in that small, barely-larger-than-a-jail-cell dorm room, not long after my summer in Boston, that one sincere prayer changed my whole future. I've always tried to get to the root of things when studying, to really grasp those concepts and ideas that are most important. I took this same approach when studying the Bible as well. God is Love. . . . For God so loved the world. . . . Jehovah, the happy God . . . over and over throughout the Bible, Jehovah is described as a loving God, a very pleasant person. So much is love a part of his nature that the Bible says that God *is* love, not *has* love. I realized from my studies that what mattered most to God, what mattered most in life, that thing that I needed to know above all else, was simply how to love and be loving. When I had that epiphany, that night, with my forehead pressed to the cold linoleum tile in my dorm room, on bended knees, I asked Jehovah in a simple and direct prayer, "Please teach me how to love." That was it. I didn't know what learning to love would actually mean, or the great suffering I would have to endure to learn how to be loving, but I knew that God could help me. And I sincerely wanted that above every other desire and goal I had.

That night, while asleep, an angel appeared to me in a dream. It was the most beautiful and awe-inspiring thing I'd ever seen still to this day. The angel was golden colored and shone so brightly that even in my dream, all I could see was the outline of a humanoid figure without any facial features. A beautiful masculine voice spoke to me, and I was filled with fear and awe. "Why do you believe?" A simple question, but one which I was totally unprepared to answer. I'd never really thought about it. The real answer, the one it would take me years to understand both intellectually and spiritually, was that I had faith based on what I'd read in God's word and the prophecies I'd seen fulfilled with my own eyes. But that's not what I said that night. "I have faith with all my heart," I blurted out without thinking about it with enough forethought. At that, the angel gradually departed from me into the distance without making a response.

I awoke suddenly—it was 6 a.m. or thereabouts—and I was saddened immensely that I'd likely responded with the wrong answer, since the angel made no response. His quiet withdrawal left me with far more questions than answers, and it was discouraging, as much as his presence in my dream was uplifting and overwhelmingly positive. I kept on wondering throughout the day what the dream meant, but I had the overwhelming feeling I'd let God down in some way with my less-than-carefully thought-out response. Still, I endured in my worship as best I could, put the dream behind me, and eventually grew spiritually to the point of baptism.

Now, I am an anointed Christian. "Anointed" might mean something different depending on the particular religious denomination an individual belongs to, but for Jehovah's Witnesses, to be one of "the anointed" is a unique honor. To be one of the 144,000 Christians who go to heaven is one of the distinguishing features of Jehovah's Witnesses, for one of the Bible's clear teachings is that God's original purpose for the Earth was for it to become a paradise gradually. We know this from Isaiah 45:18 and Ecclesiastes 1:4, among many other scriptures.

Daniel 2:44 states that God will crush and put an end to all these other kingdoms, and his kingdom will stand forever. This is the kingdom Jesus prayed for and asked to come in the model prayer in Matthew 6:9. While individuals read the Bible for various reasons and are inspired, and rightly so, by the superlative example that Jesus set, the miracles he performed had a very specific purpose. The kingdom Jesus himself prayed for is God's instrument for achieving his original purpose for humankind, namely, to transform the Earth into a paradise and return humanity to the perfection Adam lost through his sin. If Adam and Eve had remained faithful to God, there would've never been a need for the kingdom, and no anointed Christians would've ever existed.

When Adam and Eve became unfaithful, God's purpose immediately included restoring the human race to perfection through the birth of his kingdom. Instead of choosing angels to rule that kingdom, Jehovah instead chose to include 144,000 members of the human race, men and women who've faithfully endured every manner of difficulties and injustices. Such individuals he hand-chose knowing they would be both incorruptibly faithful and entirely empathetic to the plight of humankind. My path to being sealed as one of God's anointed was anything but typical. It took twenty-five years from the time of my unique dream all those years ago until relatively recently, when Jehovah revealed I was one of his anointed ones. And it proved to be just as the scriptures describe: everyone whom God receives as a son must be scourged by him.

While many anointed Christians are chosen after many decades of dedicated faithful service, my sealing was accomplished after both a period of faithful service and a shorter time when I had stopped serving God completely. For fifteen years, I was faithful through thick and thin in my worship. I had in fact maintained my sexual abstinence from college for more than fifteen years before meeting the woman who would eventually become my wife. After a period of slander and vicious gossip that originated right from within the Christian Congregation, I became so discouraged and disappointed that Jehovah could let something so sad and poisonous come from among my own spiritual brothers, I stopped serving him. Wrongly imputing improper motives to God, that he had somehow let me down, led me into a course of sin that I persisted in for seven years. Little did I know, it was during this time of unfaithfulness my true test would be accomplished, and Jehovah would use this time to know if I was worthy of adoption as one of his spirit anointed sons.

When I left Jehovah, essentially stopped serving him altogether, it was one of the most difficult times of my life. My worship had been the center of my life and all I'd known from the time I was a young man. All my close friends are Witnesses, all my goals were centered around true worship, and my hope to live forever in earthly paradise had sustained me through many very challenging times and temptations. When you lose that, the pain you feel when you truly love Jehovah is so heart-wrenching, people can seek refuge in all manner of things that can never equal God's love.

For me, at least initially, I found refuge in simply living a life devoid of the habits that characterized my worship for over fifteen years. I stopped attending Christian meetings, stopped reading the Bible, stopped praying, and stopped talking to my friends. I felt so betrayed that I wanted nothing to do with Jehovah or his people. I still loved my friends, of course, but I knew that they would try to encourage me and would be hurt by my course, so I avoided them completely. Eventually, I moved, changed my phone number, and just tried to move on with my life. I married my wife, who ironically had studied the Bible with Witnesses for years as a child, unbeknownst to me at that time, and we had a child together.

After six or seven years had passed, my life felt completely empty. I had a beautiful girl I loved, and when she almost died in utero, I begged Jehovah to spare her life, which he did. The older she got, the more I knew I needed to make an effort to teach her about the True God. In the years of my absence from serving God, I studied Buddhism and various esoteric traditions, including shamanism. I eventually realized that so-called magic was very real and began taking seriously what I was studying in various African and Native American spiritual traditions. In time, I grew to understand how very powerful these spiritual traditions were in terms of both their cosmology and spiritual values. Indigenous religious practices are, in fact, very real forms of occult technology that enable one to manipulate their physical reality. Through the use of rituals and various practices, individuals can indeed bring wealth, health, material prosperity, and other forms of advantages through the cultivation of relationships with various spirits. But I never became initiated. I never forgot what I'd learned from my study of the Bible, and I knew deep down that Jehovah strictly forbade the practice of such things. Still, I'd lost my faith and wondered if possibly I'd been wrong all those years as a Witness, and that maybe shamanism was also a path to serving God and helping others.

The more I learned, and the more I reflected on the lives of Indigenous peoples, the more I realized that my previously conceived notions about various forms of spiritism lacked nuance. In many respects, there are those within various Indigenous groups who display Biblical principles of kindness and love much more fervently than many who profess to be Christians. Still, I knew what Jehovah says, that he is the only one to whom we should pray and seek for spiritual guidance and assistance. No so-called ancestor, no other deity, should ever receive our devotion according to Jehovah's word, the Bible, and I just couldn't get past that.

I prayed to Jehovah more fervently than I had in years, and in his own way, he reminded me of the promise I'd made to him decades ago, and that I'd been unfaithful in fulfilling that promise. Psalms 119:57–59 reminded me that I'd promised to serve him, and when I read that scripture, it not only reminded me I'd made that promise, but the "why" of how I'd come to make it. I repented and gave way to tears for the first time in years. Jehovah's way is perfect, and he blesses us for simply doing that which we should do, things that are for our own benefit. The Bible's direction covers everything in life. We're never left wondering how Jehovah feels about anything. How to raise a child? How to live a morally clean life? How to treat one's marriage mate? All that and everything else that is truly essential in life is covered by the Bible's clear and principled direction. We also learn of God's future for the Earth, the wonderful promise of the resurrection of the dead, and countless other comforting aspects of God's nature and purposes for humankind.

I returned wholeheartedly to Jehovah. I took every item and book my wife and I had that related to spiritism and burned all of it. It took me a year to quit smoking, a habit I'd picked up a few years after leaving him, but I did quit successfully with his help. And when I was smoke-free for about six months, Jehovah revealed to me that I was one of his anointed. I was shocked. I couldn't believe that he would choose someone like me, someone who had been so unfaithful, someone who'd left him, intentionally. But he did. Being anointed was never anything I even wanted, never something I'd even considered before being chosen. I loved the paradise Earth, and the hope of living forever as a young healthy person and enjoying all Earth, and being a perfect human has to offer. Even now, the idea of leaving my daughter and wife on Earth while serving in heaven as one of Jehovah's chosen king-priests is not always a happy thought, but I have faith he will comfort us as I fulfill his will for my life.

People just coming to know Jehovah often ask, "How does a person know they're one of the 144,000?" That's a great question, and the Bible explains things very clearly. In Galatians, Romans, and other places, the Bible indicates that "All who are led by God's spirit" are his sons. If you don't know for sure, you are not one of the anointed. The direction Jehovah provides through his spirit when a person is anointed is so clear a person becomes absolutely sure they are of the 144,000. That certainty, however, happens for different people in different ways. For me, I knew assuredly when I read Paul's letters through the Christian scriptures that when he was speaking to those with the heavenly hope in those congregations, these scriptures also applied to me. For twenty-five years before my anointing, I'd never felt that way. And when my hope changed, I knew it definitively.

When I prayed that simple prayer all those years ago in my tiny dorm room, on bended knee and face to the ground, I had no idea Jehovah's answer to my prayer would take years of trials and painful struggles. But there are aspects of love that can only be understood fully by time and pain. Endurance, loyalty, patience, kindness, forgiveness, chastity, and other vital aspects of love can only fully be understood by enduring trials. But too, other aspects of love can only be fully grasped through other positive experiences. The ecstasy of romantic love, the superlative love a parent feels—that love which supersedes every other human-to-human love—Jehovah also helped me to know in time. But the greatest love I've known and will ever know is feeling a portion of the love Jesus feels for his God and Father. That only comes when a person is one of Jehovah's anointed ones.

Most Witnesses don't truly understand what being one of the anointed really means. It's not something you can know as a human if you're not one of the anointed. While it is a great privilege to have the heavenly hope, the true anointing, the true reward of having the hope to live forever in heaven as a king and priest lies not in the promise of the positions themselves, but in the spiritual gift Jehovah gives when he anoints one. The love of Christ is poured into one's heart (Romans 5:5). That is the true anointing, not some shallow hope of just attaining a position. But to feel a measure of what Christ feels for his father, Jehovah, that is what it truly means to be one of the anointed. Jesus' love for his father is so deep and engrossing, it forces one to want to be obedient as best they can be to anything Jehovah asks. And learning obedience to God is the true test of one's devotion to being loving.

After almost thirty years of endeavoring to serve him, Jehovah's true answer to my prayer, after years of both suffering trials and joys, came in understanding a simple but profound truth. To obey him fully is the truest way one can come to love. Jehovah is love, and not surprisingly, his laws and principles are a reflection of who he is, what he values, and what he aspires for humanity to be and become. God's laws are so perfect that when a person obeys Jehovah, they will manifest love in all their dealings, even when they may not necessarily even intend or want to. When someone wrongs us, slanders us, betrays us, the human response is often one rooted in reciprocity, to return the evil. But Jehovah's way, the path of agape, of principal love, is exalted, is higher. To show restraint, not return the evil, and obey both secular and spiritual laws, to be patient and have confidence that Jehovah will right all wrongs in his own time with his kingdom shows one's commitment to God, to agape, even when such may not be something we enjoy doing. Obedience to Jehovah's perfect law forces us to love, and when we truly love him, we will be more obedient. This is the cycle of love and obedience, and it is the most important of all principles for a Christian to understand. If we love and obey God, no matter the circumstances, with confidence he will always fulfill his promise to supply the strength that we need to do so, we will obtain his promise of life everlasting, regardless of whether our hope is heavenly or to live forever in paradise on Earth. That is what true love is. That is Jehovah's most beautiful hope and gift to all humanity. The opportunity to demonstrate our love for both him and others by obeying him intentionally, understanding that this is the highest way of living one can attain. That is what true love is. That is the gift Jehovah offers to both his angelic and human creation. What a loving Father he truly is!

A Physician Enters the World of Spirit

My wife and I went to Cuba for the third time shortly after our third wedding anniversary. Our first visit together was years before, but it had been ten years since she'd been back to Cuba, and her family hadn't seen her since she was a young woman. It was important to me that our children get to know their heritage and their many Cuban family members, at that time, previously unmet. So I encouraged her to let old wounds die. For then we could move forward as a family united and start the process of healing long, needlessly lingering wounds.

My wife was raised by her grandparents because her parents, in their late teens at the time of her birth, were both in college and didn't want to quit school to take care of a baby. It was a decision partly their own, but also one in harmony with the wishes of their parents. Both teenagers were intellectually gifted, and neither set of new grandparents wanted their children to leave school.

For years, my wife resented her parents' decision. Still, for her well-being and the sake of my wishes to help our children learn more about their heritage, she finally accepted her father's invitation to visit. He'd been trying for ten years to reconcile with her, and finally, she accepted his attempts to be a part of her life again with some assistance from me. Though she'd forgiven him and no longer held resentment toward him or the family, she'd not been back to the island in a decade. She had little interest in connecting with her family, most of whom still live in Cuba, for a number of reasons. Nevertheless, and with some initial reluctance, we began making yearly trips to the island. It was a decision that would forever change our lives.

Cuba is an easy place to fall in love with. From the time I was a child, I've traveled throughout the Caribbean, from the Bahamas to Montserrat, from Puerto Rico to St. Kitts, but I've never seen beaches more beautiful than those on the large communist island my wife called home as a child. The turquoise waters and picturesque white-sand beaches of Cuba have ruined my trips to the opaque, blue-green lakes in New Mexico where I live, which I still visit to go fishing and kayaking with my children.

Beyond its physical beauty, however, I find Cuba's true jewel to be the Cuban people. Cubans are good-humored, active, educated, and fun people. Physically, they are among the most beautiful people anywhere in the world. They come in every shade of skin color from white to caramel and dark chocolate, these colors being a product of Cuba's mixed heritage of Spanish, Indigenous, and African peoples. I know dark-skinned Cubans with gray and blue eyes, and I've seen every shade and variety of hair, from curly blond to straight jet-black like my beautiful wife's hair. Although I don't know the statistics, through observation, I would say the rate of obesity is meager compared to other countries. It's not uncommon to see muscular men and women in their sixties and seventies carrying packages of food and metallic jugs of water down the streets.

When I visited Cuba for the first time, I was overwhelmed by the degree of poverty I saw in the coastal town my wife grew up in. It's still largely an agricultural community today; the main transportation in the town is by horse and buggy. The infrastructure, particularly schools and roads, is worse in the less urban areas of the country than in the most impoverished communities I've seen and worked in throughout the United States, and I've worked in many poverty-stricken ghettos and visited several Native American reservations.

Cuba's poverty is overwhelming. It's nearly all-encompassing, and yet the Cuban people are among the most benevolent, light-hearted, well-educated, and hospitable people I've met anywhere in the world. These so-called enemies of America have been nothing but hospitable to my family and me. We have known no ill will at any point during any of our visits simply because we are Americans.

My experiences and observations have shown me that many of the things I'd been "taught" in school about Cuba and its people were incomplete and inaccurate. As I got to know the island and its people, I learned much about its culture, history, and religion, specifically about Santería.

It was the day before my father-in-law's fiftieth birthday that my first real scientifically unexplainable experience began. We'd been in Cuba for about a week, and my wife mentioned to her father that she wanted to have a cleansing ritual done. Her grandmother, who'd been a santera, had been dead for over a decade, and my wife had grown fond of another santera, Francesca, whom she'd known since she was a teenager. Francesca told her during a trip to Cuba ten years earlier that she would marry a doctor with a strong connection with the spirits and that she would have several children.

Years later, after this prediction, my wife and I met, fell in love, married, and had children. But she didn't fully appreciate Francesca's words because I had no previous experience with Santería or traditional healing and was not a man with any deeply religious sentiments. Still, Francesca had helped my wife before with other problems and told her other things that eventually came true, so she still held Francesca's words in her heart. After all, part of her prediction had already proved right: my wife had married a doctor.

My wife, thanks to her natural gifts and experiences, can feel the presence of spirits and is aware when a santera is close. After taking a walk through the downtown area where she grew up and eventually to the beach and the park, we saw a man sitting on a bench near a street that runs parallel to the beach. "He's doing something," she told me. "I can feel it." What my wife says she feels is a heaviness around her. The weight of something with a powerful aura is how she describes it to me.

We returned to her father's home to find Francesca sitting on the porch in one of her father's wooden chairs. "It wasn't him I was feeling," my wife said upon seeing her. "It was Francesca." There is a long staircase leading up to my father-in-law's porch. I was walking in front of my wife, so I saw Francesca first. She has a striking appearance with large dark eyes. She is dark-skinned, mid-fifties, with long hair in braids, about 5'5" tall, and slightly heavyset. We'd been expecting her for an hour or so, but she was delayed by people stopping her on the way to talk and ask for assistance. It took her two hours longer than expected to walk the two miles from her house because of multiple requests for her help.

When she saw me ascending the stairs, Francesca smiled and nodded her head. "I knew you were coming," she said to me. "When I heard you had married an American," she said to my wife, "I thought he would be a White man. But when I saw you two walking by the water, I saw an African man. Where are you from?" she asked me. "You have roots in two countries."

Her words didn't affect me at first. I'm roughly half African and half Irish, and my tan skin and curly hair make it obvious that I'm multiracial. But then Francesca became more specific. "You have roots in two African countries."

About a month before coming to Cuba, I'd completed a DNA test that revealed I was about half Irish, and my African heritage was split between Sierra Leone and Nigeria. Santería's roots are in Africa. Francesca said that the African roots she saw in me were so strong she could tell I was descended from two nations.

I'm not a man who is easily impressed, so I discounted her statements as coincidental. Perhaps my wife had mentioned something to her father, who'd then spoken to Francesca. My wife smiled and glanced at me, and then hugged Francesca. Then the two of them sat down in rocking chairs while I sat on a smaller wooden chair. We sat for the next several hours talking on the porch.

Francesca was speaking in Spanish to my wife, often pausing to nod and occasionally to look down. My wife sat there, slowly rocking in her chair and nodding as Francesca spoke to her. I sat quietly listening to Francesca as she began describing my wife's problems and the personalities of our children, whom she had never met. She knew and said some deeply personal things to my wife about our children, marriage, and current life concerns.

Francesca then turned to me and began speaking. Having heard her words to my wife, I listened with less skepticism than I'd brought to Cuba. At the time, my wife and I had recently had our third child, and the baby was demanding more time than our older children. I was spending less time with the older children and feeling disappointed as I realized my relationship with them had changed some. By the time of this trip, our third child had become a toddler, and, as is usually the case with a toddler, she was demanding more of our attention than her two siblings.

Francesca looked at me and then did the same routine she had done with my wife. She looked to the side or down, nodded her head, and then spoke. Her first words to me were, "You often can feel when something bad is going to happen." I nodded. My wife and I lost a child when he was young, and in the days before he died, I had an overwhelming sense of doom that his condition was beyond recovery. I'd mentioned these feelings to no one, not even my wife. His condition was serious, but there was nothing about it that made his death a certainty. Though I'd experienced such a feeling before, the sense before my son's death was immeasurably more intense and saddening. I can best describe what I felt as a sense of deep mourning inside my heart days before he died, before there had been any change in his condition.

Francesca then began describing my children to me, their personalities, things that someone who didn't know them would never know. "You need to spend more time with your son," she said. "He has everything, but he still feels sad." I knew immediately that she was right. It was something I'd been thinking to myself for some time, and yet with the demands of work and a growing family, it was difficult for me to do as many activities with him as I'd previously done before the birth of our third child.

That small piece of insight was the only real advice she gave me, but it was the most important thing I needed to hear about balancing my lifestyle. She then turned to my wife and said, "He doesn't know who he is," referring to me. I immediately felt offended. I've felt confident in who I am for many years. My wife looked at me and said, "You misunderstand. She means that you are not aware of the gift you have." Then Francesca continued, "You are very powerful, much more than me, but you do not know your gift." I looked at her, still not fully understanding, as she continued, "You sometimes see a shadow in your home in the shape of a man. He's in the corner of your eye. When you turn and look, he's gone." It was true. Again, this was something I'd never discussed with anyone, not even my wife. I didn't understand what I was seeing and didn't care to, either. I didn't believe in spirits, really, and certainly didn't give any thought to ghosts or anything that could explain from a fanciful point of view what I was seeing. It wasn't something that happened frequently, and it didn't bother me. I knew I wasn't mentally ill, so I felt no need to resolve it. "Someone with a large family owned the home before you," Francesca said. I nodded, and she looked at my wife and said, "You need to do a cleansing of a home when you move in if it's not new. When you go home, leave a bucket of water in the living room overnight. In the morning, throw it out. The spirit will then be gone."

She turned back to me. "You have a brother and a sister." Again, she was correct. "Your brother has diabetes, but it came on suddenly and unexpectedly."

"Yes," I said.

Francesca continued. "His diabetes was brought on by stress. A woman with Asian eyes consulted someone, and they did something to him. He also has a child who gets sick frequently, but it's an illness the doctors have not been able to explain. The child's illness is from *the bad eye*. When he was young, he was given a red bracelet to protect him from *the bad eye*, but when he was baptized in the church, the minister told his parents to remove the bracelet, and they did."

My brother was in his forties when he was diagnosed with diabetes. He was healthy prior to his diagnosis, and we have no family history of the disease. He wasn't obese, nor did he have any habits that could explain the illness. Days later, my wife asked my sister-in-law about the red bracelet, and if what Francesca had said about it being removed at his baptism was true. It was. It was not something either my wife or I had ever known or spoken about with my brother or his wife, so there was no way this very personal experience could ever have been shared with Francesca.

Francesca had more to say. "Your dad is dead."

"No, he's not," I replied.

"Not the man who raised you," she clarified. "Your real dad." She was right. I'd never gotten to know my biological parents—I was orphaned at birth—but the man I considered my "real" dad was actually my adoptive father. He'd raised me from the time I was a young child and treated me as his real son. But he was not my biological father. Again, this was something that, though my wife knew it, was not shared with her family, who had never seen my adoptive parents or known the truth about my childhood. "Your real father was a powerful and wealthy man," Francesca told me, "but he never spent much time with your mother."

She then raised her eyes and stared at me. "How many women do you have?"

Confused and shocked, I leaned back. "*Nada. Uno solo*," I replied, looking at my wife, afraid she would think I had somehow been unfaithful.

Francesca looked at me and smiled. "That's right," she said. "You are a good man. You've never been unfaithful to your wife." She was right again. My wife looked at me and smiled.

Francesca went on. "You have written many books, but you've not finished any of them." My wife and I both nodded in agreement. I frequently have so many ideas, I usually write thirty or forty thousand words of a novel or nonfiction project before getting bored, losing focus, and wanting to work on something new. It's a problem I've had for years but never really cared much about. I never saw writing as an actual job, just a hobby for me, a release from the stress of practicing medicine. But from my wife's point of view, it's been a bad habit because I was spending time on something that wasn't a significant source of income for the family.

My wife and I sat on her father's porch with Francesca for several hours, mostly with the two of them talking and me just listening. Francesca would nod and speak, telling us things a little at a time. She was a long-time friend of my wife and her family, and their meetings were usually a mixture of socializing and consulting. It was a custom my wife had long been used to. But as the two of them spoke, I sat ruminating on all the things Francesca had said to me, trying to come up with some logical explanation for how she could know such things about me and my family.

After she left, I stayed with my in-laws and met other friends and family members who would come to my father-in-law's house in the afternoon and evening to talk shit, drink rum, smoke cigars, and play dominoes, as is our custom while in Cuba. We stayed up late into the night that night, and all the while, my conversation with Francesca lingered in the back of my mind.

The next morning, I woke up early after a vivid dream in which I'd seen Francesca. We hadn't spoken any words to each other in the dream. In it, she appeared suddenly in my room with a young man whom I didn't recognize. I sat up in the bed, where I was sleeping alongside my wife, completely aware I was in a dream state. The young man was troubled. I could see the anguish on his face; it was a sight I'd seen many times in the faces of patients in the emergency room. Francesca simply held out her hand, and I watched the young man vomit out three spirits. She took the spirits in her hand and threw them away, and the young man immediately felt great relief. Francesca looked at me confidently and then disappeared, and I woke up.

The dream left me with an extraordinary feeling. I knew it had significance, like it was clairvoyant in some way. It was lucid, so clear. I didn't know what to make of it. I walked out of the bedroom to find my father-in-law already awake and preparing the strong Cuban coffee that usually helps us recover from our late nights of drinking and dominoes. When I told him about my dream, he nodded but seemed to think there was nothing significant about it. We drank our coffee, and after my wife and other members of the household woke up about an hour later, we all went on with our day.

My wife and Francesca had been so busy talking the day before that Francesca hadn't had time to do the cleansing ritual my wife wanted. My father-in-law called Francesca and made arrangements for the three of us to go to her home later that afternoon. It was a swelteringly hot day, so instead of walking the two miles, we took a horse and buggy.

As we approached Francesca's home, my wife began to sense the heavy feeling she usually feels when she's near a santera. She asked me if I felt anything, but I didn't feel anything ominous. I had the feeling of *something*, but it was a pleasantly familiar feeling. We were nearing the house, which was modest by Cuban standards and made of concrete with a small electric line attached to the upper right side of the porch. There was a small dog sitting on the porch. It approached me, and I rubbed its belly while we were waiting for Francesca to come to the door. She wasn't home when we arrived, but her daughter answered the door and informed us she was just a short distance from the house and was on her way. She ushered us into the living room, and we sat in wooden chairs for the next fifteen minutes talking with her daughter.

The floors in the home were covered in vinyl tile, and a curtain made from a bedsheet separated the living room from the rest of the house. By the front door stood a small plate with metal chains, two small toy guns, and some other small objects I didn't recognize. There was also a small coconut with a hole in the middle sitting next to the plate. When Francesca arrived, she lit a cigarette and placed it in the hole in the coconut. She said she often gave a cigarette to her saint as an offering in this way. On the far side of the living room were two beautiful black dolls wearing elaborate dresses. One was wearing blue, the other yellow, and there were golden crowns on their heads. The dolls represented the Virgen de la Caridad and Yemaya, two of the primary Santerían saints.

We talked for a few minutes, and then Francesca invited us into her bedroom, where her altar was located, to perform the cleansing ritual.

Every santera makes an altar for her saints. In the presence of the altar, santeras can perform more thorough consultations, more powerful rituals, and enter transition, where they channel spirits more easily. The altar increases the santera's ability to communicate with the spirits through whom all her works are achieved. As Francesca led us through her home, we passed a small bedroom on the right and then entered another room that held Francesca's altar and bed.

My eyes were immediately drawn to the altar, which was located in the far-right corner on the back wall. The altar sat on a coffee table that also had several small dolls and flowers on it. Most of the dolls were dark-skinned and female, and most were dressed in white, although some were dressed in blue and green. There were also two male dolls, one of which was larger and made of painted plaster or ceramic instead of cloth like the female dolls. It sat at the base of the altar in the middle of the coffee table. I cannot recall the other male doll clearly, but I think it was a dark-skinned figure dressed in white.

Above the altar was a small bookshelf, about three feet by six feet, with the top of the shelf touching the ten-foot ceiling. There was a bowl of cigars on one of the lower shelves of the altar, and a slim but tall bottle of clear rum sat in the middle of the coffee table in front of the altar. To the left of the altar and the coffee table was a full-sized bed.

As we entered the room, Francesca invited us to sit on three chairs located to the right of the door. There was also a small bench to the left of the chairs where we sat. We took our seats, and Francesca began placing two cords of small Christmas lights around the frame and shelves of the altar. She plugged in the lights and sat down on the edge of the bed in front of us.

Francesca next began looking at the altar, then back at us, at times pausing to nod her head. Before she began speaking, she looked at us again. I don't recall the specifics of the conversation, which was mostly focused on my wife. I was still thinking about the prior day's events and my dream. To perform the cleansing, Francesca needed some leaves and rum. Unfortunately, while en route to Francesca's home, we had forgotten the rum required for the cleansing. My wife's father called my brother-in-law and asked him to bring us some rum from the house of another relative who lived close by.

It took about thirty minutes for my brother-in-law to arrive. During this time, Francesca continued discussing various matters with my wife and father-in-law as I listened quietly. When the rum arrived, my father-in-law asked for a cup of water to be brought and said that he would be going into transition. He then stood up, and Francesca took several leaves and began waving them in front of his face and sprinkling some of the rum from her altar on his head, face, and shirt.

My father-in-law began to sweat profusely, then sat on the floor in front of Francesca. When the cup of water was brought to him, he closed his hand, extended his index finger, and began dipping his finger into the cup of water Francesca's daughter had brought into the room. He dipped his finger in the water several times and immediately began speaking in a language that wasn't Spanish and which I couldn't understand.

He then motioned for my wife to sit down in front of him. After she did so, he pointed at her right hand. She extended it, palm up, and he touched her palm with his index finger and held it there. My wife then began staring at her father's face. After about thirty seconds, my father-in-law lifted his head, exited transition, and began shaking his head and laughing. It seemed as if he was surprised by what he was experiencing. As my wife stood up and returned to her chair, her father also stood up, smiled, and continued shaking his head and laughing. My wife later explained that after her father touched her hand and she looked up toward his face, she saw San Lazaro standing behind him.

San Lazaro, or Babalu Aye, is commonly known in Santería. He is beloved by devotees and santeros alike, for, through him, many santeros have accomplished acts of healing. He is considered as the ruler over all matters related to health, both physical and spiritual. He often appears as a humble, elderly, infirm male with trembling arms and legs and sores throughout his body. It is likely for this reason that Babalu Aye was synchronized with the Catholic saint San Lazaro.

My wife didn't seem surprised to see the spirit. "I've seen him many times during rituals," she said, "and he always looks the same." This time, she only saw his face and upper torso. She saw a slim, elderly male with bright white light for hair. The light was so intense it made it difficult for her to see his face clearly, but she could tell his eyes were squinting and he had a small nose. It was after my father-in-law sat down that my brother-in-law arrived with the rum, which he gave to his father. Francesca invited him to sit down with us to be a part of the cleansing. She took the bottle of rum, gave us each a sip, and then poured some of the rum on the floor in the outline of a cross. Next, she lit the rum and had the four of us walk around the flame. As we walked, the flames took the shape of a vortex and rose about three feet. I looked down at the flame as I walked and felt the heat. I also watched as the flame repeatedly moved toward my brother-in-law and touched his arms several times. Gradually, the flames went down, and we stopped walking. Francesca, who sat on the bed as we walked, stood up once the flames had gone out. She repeated the procedure, pouring more rum in the outline of a cross and lighting it. This time the flame was blue, and she told us to walk more briskly. Again, the flames formed into a vortex, but this time the flames rose to about eight feet. Again, as we walked around the vortex flame, it followed my brother-in-law and touched him several times. But he was not burned.

My wife, her father, and I sat back down on the chairs, but my brother-in-law sat down on the small bench to our left. Then he started crying suddenly, sobbing like a child. His father walked over to console him, but instead, my brother-in-law went to the bed where Francesca was sitting. He lay down, face down, sobbing next to her. This went on for a couple of minutes until he crawled forward on the bed and put his head in Francesca's lap.

He rolled over and started laughing in a voice that wasn't his. His facial expression also changed. Francesca looked down at his head, still in her lap, and said, "That's enough." At that point, he crawled to the edge of the bed and vomited a small amount of saliva several times. He then got up off the bed and returned to the bench. He was sweating profusely as he chain-smoked several cigarettes and said to his father several times in his normal voice, "I told you something was wrong with me."

My father-in-law turned to me and said, "This was your dream." He told my dream to Francesca, who nodded and asked me if she had said anything to me in my dream. I replied truthfully that I just listened and watched, and she never talked to me. "The dream was a gift for you," she said, "for you to know the power of the saints. I don't know why it was given to you, but it is important."

Shortly after that, we left Francesca's house, and my wife, her brother, and my father-in-law all thanked me for helping her brother. When I told them I didn't feel like I had done anything, my father-in-law said, "It was because you came to Cuba that he was helped." My brother-in-law never wanted to see a santera. He was afraid of what they might say to him because he'd done things he wasn't proud of, and he knew Francesca would be able to see those actions if he ever talked to her. Even though he knew something was wrong, he was too ashamed to seek help. But she was—and still is—a friend of the family.

The next day, my wife and I went to the beach, and we didn't see her family again until the day after. Still pondering over the previous days' events and wondering how her brother was doing, I was curious as to what the experience had been like for him. The notion of going into transition and then exiting it was nothing like anything I'd ever seen. I wasn't sure what had actually happened. I couldn't explain it logically. Are spirits real? Was he possessed by a spirit before the cleansing? Did Francesca remove a spirit, or did the ritual actually cause the spirit to enter him? I ended up having more questions than answers.

I talked to my wife, who had routinely seen similar cleaning rituals since she was a child. She explained the whole process to me as she understood it. When a person has bad energy or someone has cast some sort of a curse or spell on them, shamans can clean and remove the bad energy in several ways. Some santeros can absorb bad energy directly, often simply by touching a person or praying with them. The bad energy is transferred during the cleansing ritual when objects like a raw egg or a collection of leaves are used. At times, the energy is so bad a santero will guide a spirit into a person's body and then out again. When it is inside a person's body, the spirit can remove the bad energy.

Later, it was further explained to me that my brother-in-law needed the cleansing because a santera had done something to him because of some pain he had caused her granddaughter. During the cleansing, Francesca channeled the spirit into him and out again, and in the process, the previous work the other santera had done to punish him was removed by the spirit that now entered him. For months, he had been feeling unusually moody and having problems with fatigue and difficulties in his studies at college. He knew something was wrong, but he couldn't explain it, even though his father kept assuring him nothing was wrong.

When I saw my brother-in-law a few days later and asked him how he was feeling, I could see the relief on his face. He smiled, and we went outside and talked on the porch for a while. "I thought you would see me differently after what happened at Francesca's," he said. "I don't see you any differently, my friend," I said. I then asked him to explain to me what he felt when he went into transition, when the spirit went inside him. He said, "It was like I was watching someone else inside my body. I didn't feel anything."

"But you were crying," I said.

"It wasn't me," he said. "It was the spirit. I didn't feel anything. I could see everything that was happening, but I couldn't control my body. It was like I was just watching everything happen."

I was curious. Had what Francesca did helped him or made it worse? What had occurred to him scientifically, and how? Despite my dream, despite what I'd been told about my life, despite everything else I'd seen during the day before, and everything my wife had told me throughout the years, I still had trouble processing what I was actually experiencing. *Are spirits actually real?*

In my childhood, I was taught by the Bible and movies that I should avoid anything like mediums and anything that has to do with spiritism. My experience with religion taught me that only bad spirits interacted with humans, that angels are never seen and don't interact with humankind. Now, I wasn't sure what I actually believed. Thinking about such things never had a place in my daily life. I'm a physician, and the time I've spent in study in recent years is always related to science and my work. I enjoyed all types of literature and art in the past, but my life now is always focused on my work and my family.

Religion, angels, spirits, and the like . . . they all seemed so distant to me. It takes faith to devote yourself to God and religious beliefs and live up to those beliefs. You have to spend so much time thinking about things you cannot see, and there's always so much in the physical world that demands my time and attention. I didn't deny the possibility of God and intelligent life outside the realm of the human experience, but because that wasn't something that affected how I lived my life, it didn't have much relevance for me. My family is everything to me, and I've long tried to treat others the way I want to be treated. Showing kindness and generosity have been a part of my character since I was old enough to know what kind of man I wanted to be. For me, being that kind of man is more important than participating in organized religion, although I have many friends from various faiths, whom I consider to be very religious. I've also known my fair share of religious hypocrites and pious idiots, of course, people who criticize anyone who doesn't believe like they do or see the world through the lens of their religious dogma. These individuals, though frequently well-intentioned, often fail to show basic human decency and kindness toward people they feel they are somehow superior to.

I spoke with Francesca several times over the next couple of weeks. I had many questions and wanted to try to understand, as best I could, what I'd seen and experienced. I wanted to hear her point of view. The last time we spoke, she told me, "You're not going to become a santero. You are a missionary. The saints have chosen you for something. I don't know what it is, but it is important." That was the last time I saw her before we left Cuba, but it wasn't the last time my wife spoke with her.

We had serious trouble before we left Cuba. When we arrived on the island, my wife's wallet was stolen at the airport. At the time, she was not yet an American citizen, only a permanent resident. She was still a Cuban citizen, which I preferred, as it would allow us to purchase land in Cuba and travel on the island more freely. Her US Resident card was in her wallet when it was stolen, and I was worried the Cuban airport officials wouldn't let her leave the country without it.

We didn't think much about the theft during our visit because we spent so much time with family and friends. Between drinking and going back and forth to the beach, we lost ourselves in the moment. We spoke with a member of my wife's family, who is a lawyer and professor at a law school there, and he reassured us that she would be able to leave without a problem. We had copies of her US Resident card on our phones, and all the documents were valid and up to date.

When it was time for us to leave, however, the airport officials had a different opinion. Our flight was at 8 p.m., and as is our custom when leaving Cuba, we said our long goodbyes to a house full of friends and family around four in the afternoon. The drive from my father-in-law's home to the airport takes about an hour and a half, so I made sure we had some time to spare in case there was a hiccup. When we got to the airport, the line was long. It took us an hour to make it to the check-in clerk. (There are no computer kiosks in Cuban airports to move things along faster.)

The clerk looked at my US passport, confirmed my reservation, and printed my ticket without hesitation. When it was my wife's turn, things were different. We informed the clerk of the theft of my wife's wallet and showed her both the police report that was filed and a copy of her up-to-date US Resident card. Still, the clerk hesitated. Then she picked up the phone and called one of the airport officials over to us. As we explained the situation, the official looked at us and firmly stated that the electronic copy of the card would not be enough. My wife needed her actual US Resident card or other approved travel documents. Without them, she would not be allowed to fly. I asked the clerk how we could obtain the necessary documents. She said I needed to go to the United States and bring back the travel document, or that a US Immigration office needed to fax a travel document to the Cuban airport. The alternative was that my wife could fly to Guyana, where she would be allowed to fly to the United States because of a policy that would allow her to return home without the same documentation.

I looked at my wife and could see the despair rising in her. Our children were waiting at home with their grandmother, and I had an ER shift scheduled in two days. The kids' grandmother had already taken several days off from her job to care for them, and I knew she wasn't able to take any more time off. I had a hard decision to make. Stay with my wife in Cuba and try to figure out how to get to Guyana with no money, since all we had was a little cash and Cuba doesn't accept US debit or credit cards, or get on the plane and go to a US immigration office in Miami. We couldn't even phone the United States from Cuba without FaceTime or WhatsApp, neither of which the US Immigration Office uses. With Trump closing the US Embassy in Havana, the only choice for us to stay together seemed to be for us to go to Guyana. And that wouldn't be easy. I couldn't access my US bank account from Cuba, and I only had about $3oo in cash left in my wallet, not enough to get us to Guyana. It would take at least another day to get our family to send us money, and even if they did, I wasn't confident we would get back in time for my ER shift. I also wasn't sure that even if we went to Guyana, my wife would be able to fly to the United States.

I realized I could do more in Miami to help her get home than if I stayed in Cuba, so I reached in my pocket and gave the $3oo of cash to my wife, then embraced her and got on the plane to Miami. It was a hard decision, but I knew it was the right one. My wife was stuck in Cuba. I was told it would take at least a week or two to get the travel documents from the immigration office, and we didn't know what we were going to do for child care.

I called my father-in-law to ask him to pick up my wife at the airport, but there was no answer. He'd left the airport about an hour earlier, and now I worried about how she would get back to his home. Taxis are seldom available at the airports in Cuba unless you've made prior arrangements. Lacking a phone that could call without an app, I was worried about my wife's safety. Still, I knew we didn't have a choice.

When I got to Miami, it was 9 p.m. on a Saturday. I tried to reach my wife as soon as we landed, but she wasn't online with FaceTime. She would only be able to use the internet at her father's house, and when I landed, it hadn't been enough time for her to make it back there. I took an Uber to a hotel in South Beach and started searching the internet for what else I could do to get her back stateside. I found a resource that said the Cuban airport would sometimes accept a letter from an airline and grant a person permission to travel if the airline agrees. I also found the application for the travel documents on the US Immigration Office website and realized it wouldn't be too difficult to complete. But it cost almost $600 and would still take several days. Plus, I wasn't sure if I could complete the form without my wife being present.

I was finally able to talk to my wife when she made it back to her father's house. Her father was on the porch, and as soon as he saw her, he put his head in his hands and started sobbing. That night, family and friends came to his house to see her and offer comfort, but she wasn't in the mood for company and just sat in the bedroom.

"I just stood on the side of the road when I got to my father's home," she told me over FaceTime from her father's porch. "I was too angry to cry." My anger over her experience made me regret getting on the plane, yet I was grateful she safely made it to her father's house. I went to bed about midnight, determined to do all I could to get my wife home as soon as possible.

I woke up at five the next morning. I hadn't slept well, worrying about what we could do to expedite things. I went to the airport early. I wasn't sure where the Immigration Office was, so I went to the airline check-in first, since there was a possibility the airline might be able to send a letter to the Cuban airport granting my wife permission to travel. The United Airlines ticket agent was beyond helpful. When I told him about our situation, he immediately spoke to his supervisor, who told me they do sometimes send the letter we needed to the Cuban airport. Still, it usually takes at least a week to get things moving. The agent then changed my return ticket to New Mexico to the next morning without charge and talked again to his supervisor. Then he told me the US Customs and Border Protection Office might be able to make something happen faster than the airline could.

Next, the agent walked me through the airport to the Customs and Border Protection Office upstairs and showed me the black phone I needed to use to talk to the appropriate officer. The kindness he showed me made me feel a little more encouraged that I might be able to get my wife home faster than I thought.

I picked up the phone, and after about a minute, an officer picked up. "How can I help you?" He sounded Hispanic, which made me think he would sympathize more with our problem. When I explained the situation, he seemed puzzled. "The airline said we would be able to do something?" he asked me. I said yes. "I'm not able to help you," he replied. "You need to go to the Global Entry Office downstairs. They can help you."

I made my way back down the stairs and walked halfway across the airport to the far end of the first floor where the Global Entry Office is located. I walked into the office and sat down in one of a dozen or so chairs lining the wall in the lobby. A large, opaque glass wall separated the lobby from the cubicles occupied by half a dozen officers. I was the only one waiting. I could hear a female officer talking hurriedly to a Hispanic man. When she ushered him out, I could see from his expression that she'd been less than helpful. After she opened the door for him to exit the cubicle area, she walked into the lobby and motioned for me to come over to her.

She was a White woman in her thirties, close to six feet tall. I could see in her face that she was irritated at me for even being there.

She waved me over and asked, "What do you need?" As I explained the situation, she looked puzzled. "Customs and Border Protection told you this office could help you?" she said. I said yes, and she said, "We don't do that here. You have to go to an Immigration Office to get that done."

"But," I said, "it doesn't open till tomorrow morning. The customs officer said if you gave me proof of my wife's US residency, they could fax it to the Cuban airport, and she could take the flight today."

"That's not something we usually do," she said.

I calmed myself down and spoke in a kind tone. "You can use the computer to confirm that her US residency is legitimate," I explained, "and print out the paperwork. Right?"

"Yes," she said, "but we don't do that."

I couldn't keep the sarcasm out of my voice as I said, "Okay, thank you for your help." I turned my back and left the office. I was annoyed, watching my tax dollars being wasted on a useless officer. She could have helped me easily. There was no one waiting, and she couldn't take five minutes to help someone in a difficult situation?

I'd done everything I could at the Miami airport. The Immigration Office wouldn't be open for another day, and now I'd have to change my flight again and start making arrangements to have my shifts in the ER covered for the next several days. I kept telling myself, it wasn't the end of the world. Sure, the hospital would have trouble finding someone on such short notice to cover my shifts, but I was sure they'd succeed—they'd have to. In the ER, doctors don't call out sick unless there is a catastrophic problem. In ten years, I've only called out sick twice, first at the birth of my youngest daughter and, second, at the loss of our child. Missing my shifts in the ER just wasn't something I did. I also figured my wife would be all right, as she was with her family. Once I got to the Immigration Office, I would likely be able to get the paperwork done, though I still had mixed feelings about it solving our problems. I messaged my wife about what I'd done—or rather not been able to do—at the Miami airport, and she replied that she was going to go back to the airport in the afternoon to see if they would let her on the plane.

It seemed like a waste of time to me. I'd talked to the airline, to Customs and Border Protection, and to Global Entry, and they'd all said essentially the same thing: "Go to the Immigration Office tomorrow." So I sent another text to my wife and told her to stay at her father's house, and in the morning, I would go to try to get the paperwork done. But she ignored me.

Early that morning, Francesca was walking to the hospital in Cuba to see her daughter, who'd been hospitalized for pneumonia. On the way, she saw my wife walking with her backpack on. She walked around a corner, and when Francesca passed that corner, my wife was gone! It wasn't actually my wife that Francesca saw, but an image of her. Francesca knew immediately that something was wrong and went to my father-in-law's house to see what happened. When she arrived, my wife and her father were sitting on the porch. They told Francesca what had happened, and she told them she could help.

She told my wife she'd go home to get a few things and come back shortly. When she came back, Francesca took my wife into the bedroom where we'd been sleeping and told her to undress to her undergarments. She then had a glass of water placed on the floor and took out some leaves and a bottle of rum. She waved the leaves in front of my wife and then went into transition in front of her. A spirit revealed to her that my wife's aunt, who also works with spirits, did something to harm her because she felt wronged we hadn't given her money or spent enough time with her on previous visits. The aunt wanted my wife to have such a bad experience in Cuba that she would never want to come back. Francesca then told my wife that they needed to go to the beach to do a different cleansing ritual. But my wife said she didn't have time because she needed to go back to the airport to try to see if they would let her get on the evening flight to Miami. Francesca said she understood, but added that my wife needed to give her some of her clothes so that she could use them in the ritual on the beach.

My wife gave her one of her favorite pairs of Nike shorts and a shirt, and then she left with my father-in-law to go to a Cuban immigration office and then on to the airport. The Cuban immigration office told her exactly what I'd been told in Miami. There was nothing that could be done without the proper travel documents. Still, my wife insisted that her father take her to the airport.

While my wife and her father were traveling between the Cuban immigration office and the airport, Francesca completed the ritual at the beach. The spirits revealed to her that my wife would be allowed to travel, but they also told Francesca to walk on her knees in the sand until they bled. Francesca carried out the spirits' direction, then called my wife to let her know that everything would be okay, and she would be allowed to travel.

The flight was at eight o'clock. Around five, I began to have a calm feeling that everything was going to work out and my wife would be allowed on the plane that evening. Later, when she arrived in Miami, she told me what Francesca had said to her and had done. I still had my doubts, but with everything I'd seen and the sense of calm that came over me around 5 p.m., I believed what Francesca did had helped us.

That whole day, while our family prayed, my wife's relative, the lawyer, made phone calls. Francesca took my wife's clothes to the beach to do a ritual. Whatever it was that actually worked, that evening, my wife was allowed on the plane to Miami.

We were both overjoyed when she landed in Miami. Between everything her family did, I asked her what she thought made the difference, and she said it was what Francesca did. When my wife landed in Miami, the Customs and Border Protection Officer said that in ten years, he'd only once seen someone allowed to travel without documents. He asked my wife, "Who did you talk to, one of the ladies that . . . *hahaha,*" he laughed, shaking his shoulders, imitating a santera. My wife laughed and said, "Something like that."

That evening, I watched my wife exit the Miami airport baggage claim area, and I embraced her tightly. We took an Uber to the hotel

I'd stayed in the night before, got up early the next morning, and got on the 6 a.m. flight to New Mexico. During that flight, I sat thinking about everything I'd experienced. I had an overwhelming sense that I needed to know more. I wanted to talk to other santeros, curanderos, shamans, etc., and to people who'd consulted them. I knew if there were people who could actually do these sorts of amazing things, there had to be people with their own experiences. I needed to find a plausible explanation for what I'd witnessed. Getting the input of others was the only place I could start, and so I began my research and work.

Since the time of my first experiences with a santera, I've sought out not just shamans and people helped by them but also people from other cultures, including curanderos, witches, and other healers and mediums. I also sought out nurses, physicians, and other educated, professional people who've witnessed things similar to what I've seen and experienced. I knew such accounts would likely be received with ridicule and skepticism by some. Still, to my surprise, my quest led me to countless people within the medical community and elsewhere who've had personal experiences or eyewitness observations that verified what the mediums and healers I've met have told me.

I've included in the pages that follow several of the accounts that have been shared with me. Additionally, a significant portion of this book is dedicated to what I think are important implications of what these unique mediums and healers are able to do with the assistance of what they describe as spirit beings. Considering the implications of shamans' abilities is as important to me as the actual experiences I've been told about. We live in a world in which santeros, paleros, witches, shamans, and mediums are often ignorantly portrayed and shabbily treated. Also, many scientifically minded individuals do not even acknowledge that unseen beings actually exist. Do santeros, mediums, witches, and shamans have a place in the modern world? Are intelligent unseen beings real? Though not previously considered as such by society in general, if real, unseen entities actually represent alien life, what are the ramifications of that for society? Is there a role in modern medicine for santeros, paleros, witches, shamans, curanderas, and others who interact with these beings?

These are just a few of the questions I haven't stopped considering since my introduction to the realities of unseen intelligent beings. These questions are not trivial. They have considerable implications for society. Should esoteric and indigenous healing practices be studied for their potential usefulness to society? It is an ethical question revolting to some, and yet equally valid for consideration.

From Star Wars to X-Men — The Media and Magic

"The world is full of magical things patiently waiting for our wits to grow sharper."
—William Butler Yeats

"Any sufficiently advanced technology is indistinguishable from magic."
—Arthur C. Clarke

Maybe you're a psychic? What does it even mean to be clairvoyant? Should science even consider the possibility that clairvoyance is an actual trait/ability? Why? Why not? Different cultures call it different names. In America, "psychic" is the more commonly used term. For cultures that embrace shamanism, it's often said a person has "the sacred gift," or in Cuba, simply "the gift." Yes, different cultures call it different names, but regardless of the title applied, they essentially describe the same thing. There are some people who just know stuff they shouldn't know without any logical explanation.

Over the years, the unique ability of shamans and others with the ability to interact with unseen entities was intentionally hidden, and for good reason. Indigenous peoples faced enslavement and all manner of atrocities without their persecutors knowing the full capability of their unique technology and understanding of unseen entities. Just imagine what further evil atrocities would've transpired should the vicious individuals who perpetrated gross injustices against indigenous peoples also understood indigenous technology.

For many cultures, what can be achieved through the interaction of unseen entities was considered a sacred trust, not to be abused in deviant and selfish ways. But in time, other derivatives of Indigenous technology developed, including ceremonial magic and what some would describe as dark witchcraft. This knowledge spread to include a larger and more diverse array of cultures, but still, such knowledge was mostly hidden from the general public. Over time, though hidden, the influence of Indigenous technology in various forms became apparent within the writings of fiction authors and, in more recent times, movies and television.

I was born in the late seventies, and my older brother eight years earlier. Among my earliest childhood memories was seeing the Star Wars sheets on his bed and his Star Wars lunchbox. There were action figures scattered all around his room, but my personal favorites of all his toys, his recon trooper on a speeder bike and an X-wing fighter, sat resolutely perched on his dresser. I would serpentinely sneak into his room and fly them around the den in my hands whenever he was away. I was just a little kid when I first caught a glimpse of the phenomenon that Star Wars became, but I knew enough to know that there was something special about it. But what was it exactly? What made Star Wars connect with so many fans?

George Lucas, like many writers and storytellers, was heavily influenced by the work of Joseph Campbell. *The Hero with a Thousand Faces*, Campbell's masterpiece, highlighted the pattern common to popular myths that have stood the test of time, a pattern that is present and has persisted throughout various cultures around the world. The story of Star Wars follows the pattern outlined by Campbell nearly exactly, as have many other modern cinematic stories. But Star Wars' success is rooted in something deeper, for it nourishes an aspect of the human experience that is universal, though seldom openly discussed.

Within Star Wars is a pattern of subconscious spiritual quests that reflect aspects of various beliefs common among the world's religions and various Indigenous philosophical perspectives. It blends these aspects of religious iconography with aspects of martial arts philosophy and symbology and what is commonly thought of as magic. It is this appeal to the spiritual journey of man, a quest for the personal discovery of humanity's innate religious yearning and clairvoyant potential, that compels the allegiance of its innumerable fans, albeit in many ways primarily on a subconscious level, for few look at movies hoping to gain some spiritual awareness. It is this subconscious appeal, filled with the lore of magic, spirituality, and the martial arts, that binds Star Wars to its audience and establishes its cross-cultural relevance. Let's consider some of its symbols and themes for a moment.

The Force

The idea of a unifying force present in all living things is not a concept that originated in the mind of George Lucas. For Christians, the Holy Spirit is the binding force that gave life to humans according to the Bible. For Asians, the qi, or life force, flows through all living beings. These are concepts intrinsic to many world religions and cultures. For many Indigenous cultures, all things in the natural world, both inanimate and animate, are bound together by a sacred universal energy. It is this common Indigenous perspective that most resembles the idea of the Force Lucas describes in his masterpiece.

In addition to the obvious mystical nod Lucas makes with the Force, there is other familiar religious iconography present throughout the series. The robes of the Jedi, reminiscent of Catholic priests, Buddhist monks, and practitioners of ceremonial magic, pay homage to these ancient practices. The Jedi, armed with a lightsaber, a symbol reminiscent not only in form of the samurai's sword, embrace philosophical traditions with direct similarity to the Japanese culture of samurai. The notion of a noble and honorable warrior priest is a pattern also present in the Shaolin monk tradition. We see these symbolic and metaphorical similarities, which upon consideration become obvious and have been long noted by many observers before me. But these symbols are just the seasoning of the universal appeal of the spiritual journey of the story's protagonist, Luke Skywalker. Skywalker's journey to embrace his power, his magic as it were, and his prowess as a warrior remains present throughout the story below the surface of Lucas's story guided by the pattern of ancient mythological structure as described by Campbell.

In other written works, I've described the recurrent pattern of malevolent examples of unseen entities typically portrayed in movies and television. Usually within the context of paradigms defined through the lens of religious dogma, such works paint an incomplete picture framed through the lens of Western religious ideology's view of the paranormal.

The presence of themes centered around magic and the paranormal in modern entertainment continues to increase, often described in the metaphorical representation of magic and unseen entities occurring throughout various popular stories of fantasy and science fiction. In more recent times in cinema, there has been a significant emphasis on overt references to both unseen entities and the fictionalized power of psychics, mediums, shamans, witches, and others who have the ability to interact with these entities. This new generation of stories reveals an obvious and well-hidden fact, to which most of the general public are oblivious. There is a large segment of powerfully influential individuals within the cinematic world who understand the reality of what is often described as magic, and they have continually inundated society with these stories. In times past, a veil remained in place, and magic was assigned to the realm of children's stories like Disney princesses and heroines, which reflected in symbolic ways the realities many individuals knew but kept hidden, yet subtly represented in fiction and fantasy. In recent times, the awareness and reality of unseen entities, and what people describe with antiquated terms like "magic" and "psychic," have begun to take on more complex and accurate examples in modern storytelling.

Netflix and other streaming services, with shows like *Diableros* and *Stranger Things*, movies like *Dr. Sleep*, and countless fictionalized stories of magical heroes like the X-Men and other superheroes have subtly been revealing to humanity a reality understood for thousands of years by Indigenous people. This known but hidden reality persisted with humankind's ignorance, with the exception of a small segment of humanity.

Consider *His Dark Materials*, a relatively new series appearing on HBO Max. *His Dark Materials* presents a spectacular tale that reflects in many symbolic terms many perspectives and actualities common in occult practices. The central heroine, a girl aided by a mechanical device that reveals the answer to any question she asks, is a not-so-subtle nod to divination. Her device, which answers her questions through three symbols that can have multiple interpretations, is a thinly veiled symbol for tarot cards and other forms of commonly used instruments in the realm of the occult. The Magisterium, who don robes and advocate an austere lifestyle, is an obvious symbol for the Catholic Church and has similarly been at war with witches for generations in the show. It's a fascinating show with lots of unique and creative plot lines, but as someone who understands the reality of unseen entities and their various abilities manifested through individuals, it's easy to see through the veil of the show's symbols and characters.

All the central heroes and villains depicted in X-Men also reveal a pattern of abilities shamans and others manifest as a result of their ability to interact with unseen entities. Professor X and Jean Grey exemplify clairvoyant empathic ability; Mystique and Nightcrawler, shapeshifting and clandestine travel ability; Colossus, the epitome of physical strength and skin impervious to injury; Wolverine, superhuman healing; and Cyclops, the metaphorical representation of the bad eye. Each of these characters depicts fundamental abilities and concepts represented in virtually every culture that has a form of shamanism.

While writing *The World Hidden*, it wasn't until I approached the conclusion of writing the book that I gave significant thought to whether or not the book should be written and shared. Before that, I'd felt a personal impetus to share what I'd seen and understood for the benefit of humanity. As I understood what the ramifications of the awareness of TOTI (the technology of the Indigenous) meant for society, I questioned considerably the ethics of sharing what had become obvious to me.

It is the nature of human beings to use every technology in destructive ways, and what is often referred to as magic is no different. For centuries, the shamans of various Indigenous cultures used the ability to interact with unseen entities primarily to bring harmony and prosperity to their communities. Within *The World Hidden*, I purposely withheld elements of the work of shamans and others that I could not confirm or had not observed. Many of the more incredible abilities reported in historical anecdotal accounts stretch the belief of what is possible, and I didn't want to introduce any more inclination toward skepticism about a reality that is already difficult to accept.

The ability to interact with unseen entities, sometimes even without the understanding of the interacting person, represents an enormous advantage for both the practitioner and the individuals who manifest such ability. The various aspects of such activity, including healing and protection, certainly have merits, but the ability to perceive the future—the best path for an individual to take, for example—represents an unquestionable advantage for individuals who avail themselves of this ability of shamans. In times past, this knowledge was principally part of Indigenous cultures in the form of shamans. In more modern times, what is known as ceremonial magic developed, and the use of unseen entities became a part of various Western mystical traditions.

The obliteration of countless cultures through colonization and the Native American genocide gradually placed the use of this technology in the hands of those who principally used their ability to interact with unseen entities for selfish reasons, targeted for their own individual prosperity or the prosperity of a small segment of the population.

For centuries, the awareness of unseen entities was hidden by this small segment of the population and followed the pattern of much of humanity's technological advancement, with the focus disconnected from one engendering harmony, balance, and connection with nature to one of exploitation and manipulation of others to the harm of society. The practice of such works is often done in the dark and in secret, but its effects are clear and very real.

As I weighed the pros and cons of publishing my first book, I realized that in practice, the benefits of acknowledging unseen entities far outweighed the risks. I came to this conclusion for several reasons. Consideration of unseen entities reveals a truth that has always existed. In revealing the reality of this form of alien life, I was simply telling the truth, nothing more. The actuality of the existence of these beings forces humanity to address a deeper, more important truth, namely, what is the origination of these beings? There is a relationship to the occult that cannot be escaped when considering the existence of unseen entities, and what shamans and others are able to accomplish through their interaction with them. That relationship is either connected to some form of divine being or it is not, and answering that dilemma lies at the heart of one of the most important quests humanity can pursue.

Whatever the answer is, humanity will have taken its most significant step in the scientific establishment of God's existence or nonexistence, and that is a pursuit worth risking the potential damage that may come from the use of TOTI for harmful activity. I also considered how individuals, much like myself, born with abilities they do not understand how they should use, could benefit from understanding that they are not alone in the world, and that there are those who can assist them. I believe firmly that there are a great many erroneously diagnosed schizophrenics who are simply misunderstood mediums and shamans, born with the gift to see and interact with unseen entities.

In the world of TOTI, it is my opinion, garnered from personal experience, study, and interviews with numerous shamans from various cultures, that a person must choose one path or another. They cannot both harm and heal, protect and attack. They are either of the dark or of the light, malevolent or benevolent, Jedi or Sith, as it were. I feel a deep compassion for the segment of humanity who, much like I did, have little understanding of the gifts they have, or the ways in which they can be developed and used for the benefit of others. These individuals are a part of the family of humankind, and as such, some accommodation for their training to beneficially use their abilities should be established.

The darkness has always existed, and those with malevolent intent for using their abilities will always exist. I don't fully understand all the ways of unseen entities, but if there is some connection with the divine, there will always be a limit to what individuals are permitted to pursue with darkness. If these unseen entities have no such connection with a divine being, entities inclined toward benevolence will no doubt provide the guidance and support needed to ensure that there will always be a segment of humanity dedicated to the moral and ethical use of TOTI. It is with confidence impelled by this understanding, made firm by the guidance and support of the unseen guardians who have protected and supported me my whole life, I feel no fear in sharing the reality of both the beauty and the darkness in the world of TOTI and all its potential represents.

A Cuban Curandera Shares Her Journey to Healing

A retired elementary school teacher now in her seventies, Señora Josefa made it clear to me when I asked her if she wanted to be anonymous after her interview: "Use my name. I want people to know who I am." A beloved santera in her community in Antilla, Cuba, her experiences as a child mirror those of many healers who work as curanderas. A descendant of a healer and raised in the home of a practicing curandera, she saw the power of unseen entities from an early age. As you read her account, notice how it parallels the accounts of others recorded within this text. It was the similarities in accounts like hers as a Cuban Curandera to others, including a Nigerian Babalawo, a Native American shaman, and others from various parts of the world, that helped me to understand there must be something objective these individuals are experiencing which can be measured given the enormous subjective similarities of their experiences. I thought, "Why should individuals from very different cultures and ethnicities, and vastly different geographic locations, share the same subjective experiences? Is that not itself a sign what they are describing must have a basis in the quantifiable and objective physiological functioning of their brains?" Decide for yourself.

I was born in 1952 in a hamlet called Cayo Ceuta in Antilla Municipality, Holguin Province. Before my birth, a rail line was laid that divided Cayo into two parts. It was here I passed the years of my childhood and part of my adolescence.

I would bathe in the waters of the Bay of Nipe. When I was little, I observed my mother, who was not recognized as a santera but as a curandera. She had no altar and only one saint, the image of the Virgin of Charity. My mother told me that her spiritual work was a gift. She worked with one glass of water and helped many people with all sorts of problems, especially those with health difficulties. She was their doctor for physical health and the soul. She would work from dawn to dusk if she had to do intense work for a difficult case.

As far as I know, this gift is hereditary. When I was seven or eight years old, I was walking along a trail in my dad's small field and saw a dark-skinned man with a machete in his hand. I knew right away from his appearance that he was not a real person. I got scared, but then he spoke to me: "You sure want to take a honeycomb." I said to myself, "Ay, he's right! My God, I do not want to see any of this."

That image has stayed with me all my life. From that moment on, I had revelations in my dreams, and when I was awake, I saw visions of people and animals who looked real. It took me years to become comfortable seeing these things. The years passed, and I finished college and graduated as an elementary school teacher. But I had no peace of mind, either in my work or personally. I always tried to reject the idea that I actually saw spirits.

When my youngest daughter was born, the situation worsened. I asked for leave at school because my daughter was frequently sick. When it seemed she was getting better, another fever would appear. I told my mother what was happening, and about my life with the spirits, and how the visions were a real problem. She looked at me and told me directly, "Until you attend to those beings that you bring, you will continue the same or worse."

"I just want to be left in peace," I answered. "I do not want any of that."

When my daughter was two years old, she began having severe ear infections. She was admitted to a pediatric hospital, but the infection continued. On the eve of San Lazaro's birthday, I was sitting on my bed and felt a moment of intense ecstasy. I saw a figure with bare feet and crossed legs, and then I looked up and saw his face. He had long gray hair that reached to his shoulders. He spoke to me. "If you want your daughter to heal, give her two drops of alcohol in her ear every day. Also, that nephew of yours who has problems with the law, he will be acquitted."

At the end of a week of the treatment the spirit told me to give my daughter, a large amount of impurities came out of her ear. My nephew, who I had no idea was having any problems, was found to be without fault in a trial. After I saw what the spirit had told me came true, I started on this path. From there, I started doing what the spirits asked. I saw them as real people.

One day, I said to my husband, "I want you to make an altar with four steps." On the top step, I drew a sun with its rays, and there I placed the Virgin of Charity and Jesus. On the second step, I placed statues of Santa Barbara and San Miguel the Archangel and a couple figures of Indians. On the third step, I set African images. And on the fourth step, I set a picture of the warriors and San Lazaro. I added the offerings they asked me to give them.

Without knowing anything about me, people now began approaching me and coming to my house in search of clarity. "I do not know anything about that," I told them. "I don't know how to help you."

"But you have an altar," they responded. I felt compelled to take them into the room with my altar, and after having a conversation with a spirit, I could tell them things that would help them resolve their problems.

When I started putting glasses of water on the altar, the energy flowed and guided me with information about any problem that people brought to me. I could feel the spirits' presence everywhere

I went. When I worked with people, I always began by praying, and chants flowed from me like springs, and I went into a trance without even trying. I cannot explain why this happened. When I gave consultations using the glass of water, I almost never remembered what I'd said, and when people told me later what I'd said to them while in a trance, I always protested, "I did not say that."

I became what is called a curandera. As my reputation grew and people began visiting me more and more often, my husband began to disapprove of what I was doing. He told me he was not opposed to my helping our household with my gift, but he did not want so many strangers and people we hardly knew coming to our house. It took some time, but his attitude eventually changed.

On one occasion, my older sister was invited to a mass for a deceased person and asked me to accompany her. A few days earlier, my husband had left for Manzanillo for work. I did not know when he would return, but he came home around the time of the mass and went to look for me at the Catholic church. They told me later that my husband went into the mass and saw me in a trance and speaking in a different language. After seeing that, my husband always approached me in a loving way and never objected again.

I never had any godparents to teach me. The saints taught me everything I needed to know for my protection and for the work I do. I have become known outside of my municipality, and many people starting in the work I do have come to see me as their godmother for the assistance I give them.

Only a small part of society understands the nature of spirits and the work I can do. Many people who practice other religions do not accept my work. Sometimes they refer to curanderas and others who work with spirits as diabolical beings. However, I always say that God is omnipotent and that he does not give humans these gifts unless it is his choice. He uses his power and wisdom to serve us all and entrusts us with the path of good. I use my work to heal and help others. I do not use it for evil purposes.

When I became a curandera, I stopped teaching for five years. When my daughter was eight years old, I wanted to return to teaching and introduced myself to the head of education in my municipality. He asked me, "Is it true that you are now a santera or curandera?"

"Yes," I told him, "and you can accept me as I am or I will not continue in my healing work."

He told me, "I have to verify things with the province." The next day, he called me and said it was not a problem, and I could start working as a teacher again.

When I neglect the indications the spirits give me, I feel negative energy. But I have received many benefits from them, and now I know I am an instrument of God. I know this for several reasons:

1. I meditate and do what I can to help others with their thinking and the problems that overwhelm them. Their problems are often serious. I remind them of the power of God's hand and then encourage them to improve or resolve their difficulties.
2. I do not think that miracles exist. We receive what we deserve for our way of acting and thinking toward our neighbors.
3. My work as a curandera gives me the ability to understand and heal the pains of others.
4. I understand the ways of God to help and comfort all in any adversity.
5. The work I do takes self-sacrifice. I work for free and ask for nothing material in return.
6. I enjoy the work I do despite its difficulty and the material deficiencies I experience because God gives us the necessary energy to overcome.
7. The work gives me spiritual fortification to face the difficulties in life. When I say "Jehovah" or "Elegua" or name a saint, the negative energy dissipates.
8. I always invoke God's name and his son Jesus before consulting or doing rituals with the saints, also in fulfillment of

promises for sanitation, cleaning, etc. I always say, "It is not my work on Earth, it is yours. I trust in you, Heavenly Father," and thus, I feel close to God both within the process of helping others and outside of it. I feel and see God's communication with me and his responses to my prayers.

People call upon saints and assistance from spirits for a variety of reasons. It's common for people with problems at work, people desiring a better salary, individuals with conflict with their bosses, couples with difficulties in marriage and relationships, individuals desiring official documents for travel, individuals with health problems, and people with negative energy to come to me. That is to say, there is no end to the number of situations and challenges people face that I and others who do the same work I do can help them with. There are also those who want to see the results immediately, and sometimes they do. But for others, their path is so thorny that it must be cleansed first.

In my case, the spirits allow me to describe an individual's path clearly. I can tell them if the action or favor they seek will be given quickly or at some time in the future, and whether or not it is appropriate to take some action. Sometimes individuals get angry when the response is not what they want, but I do not encourage anyone to do something without a positive result.

I think that curses are like the processions that return to the darkness from whence they came. You do something wrong, and you will face the consequences. People need to foresee the consequences of their actions and be sensible about what they do. Laughing or making fun of someone or something they do not like or understand always leads to bad results.

Once a neighbor stole a bunch of bananas from an unhappy old man for whom I had great affection. A saint or San Lazaro (I can't recall specifically) showed me a revelation of the thief. I ran to alert the old man, but I was too late to stop the theft, so I just told the old man, "You're going to know who stole the bananas." He doubted what I told him, so I added, "For stealing your bananas, the young man will be sick and go to the hospital, and finally he will come to me, and you will be present with me when he does."

Soon the young thief became very weak and pale, and when he came to me for help, the old man was present with me. The thief told me he felt very sick, but the doctors did not know what was wrong with him. I said, "You're going to get well, but first, you must bring a bunch of bananas to this man because what you ate was not yours." The young man replaced the bananas he stole, and thus, he healed.

Another time, a man stole an offering of rum given to San Lazaro. He came to me and said, "Look at me. I have sores all over my body. I'm taking antibiotics, but they are not going away." I told him they weren't going to go away unless he took a bath with some branches from a wild tree and put back the rum he took that his mother had given to San Lazaro. He said, "I asked permission before I took it," and I answered, "But did he give you permission!" He was embarrassed and said, "He did not."

A Sports Medicine Physician Witnesses the Power of Faith

An enormous debt of gratitude will forever be owed to all the interviewees and others I've worked with to complete the compilation of this work. Chief among them is my father-in-law, who helped me obtain Alejandro's* account below and those of several other Cuban contributors who permitted me to interview them. When gathering interviewees, I focused primarily on those who had public experiences that could be verified by multiple witness accounts; however, I did not turn anyone away who I thought had a legitimate experience even if parts of it could not be verified. Various parts of the account below were observed by several members of our family and friends in Cuba.

I grew up in Cuba in the 1960s shortly after the revolution, in a town with approximately five thousand residents. There were no cars or motorcycles in town, and the nearest hospital was in Santiago, about a six-hour ride by horse and buggy, which my family didn't own. There were no doctors in my town, so nearly everyone around me went to santeros for medical care.

Santeros and santeras were common in my family and community. My mother and grandmother were santeras, but my mother was not very powerful and only practiced to protect our family and help us when we were sick. When I was young, she was not very active, and I can remember seeing her in transition only a few times when my grandmother and other devotees and santeras would be celebrating *bembé* in our backyard.

December 17 is an important day in Santería. It is the day of San Lazaro, to whom my grandmother and many other santeras are devoted. People dance wearing sackcloth, sing praises, smoke cigars, drink rum, eat sweets, and often slaughter goats and chickens to drink their blood and remove their hearts. When I was young, my grandmother participated in these rituals. I would tell her I liked the smell of cigars and rum, and she would say, "You have been chosen by San Lazaro." This is a common saying in Cuba because children do not usually gravitate to cigars and rum.

I was seven when I had the first experience that I can remember with the saints. I was lying in bed at night when I felt a stiff, cold breeze move through my room. I could feel the presence of someone the same way a person feels someone looking at them, but when I turned to look for my mother, she was not there. I felt afraid and turned away, but then I felt the breeze again. It felt like a hand touching my face. Then I felt a sensation of ecstasy and happiness all over my body. The breeze blew strong and cold again, and then I felt the spirit leave the room.

When I was young, I developed fevers and stomach pains, at times with nausea and vomiting. My mother always took me to a santera, who would perform a cleansing ritual, and my symptoms would instantly improve.

I had a much more serious problem at age eleven. That was when I saw that my mother had power and that the power of the saints was real. I was playing outside without shoes and stepped on a nail. I was afraid to say anything to my mother because we didn't have any money for doctors, and I didn't want to get into trouble. But my foot began to hurt badly. The pain got worse and worse over the next week, and by the end of two weeks, I could not move that entire side of my body because my foot was infected. I was diagnosed by a doctor in a town about twenty kilometers away with tetanus and was told I would never walk again.

It was a difficult ride home. But when we returned to our small apartment, my mother came into my room. She put her head in my lap and went into transition, then started singing a strange song in another language in a voice that wasn't hers. The expression on her face changed so much that she looked like someone else. She then began praying over me. She started doing this every day while my father was out of the house and at work. Gradually, I got better, and after six months, I was able to walk and run normally again.

At that time, my father hated my mother's involvement in Santería. He was an atheist, and Santería was illegal in Cuba at the time, though it was still heavily practiced in secret. Only Christian denominations, including Catholicism, were legal then. My father loved Fidel. He was a soldier in the Revolutionary Army and did not want my mother involved in anything that could hurt our family or affect his position in the government. My father was the president of the local province and knew that if people became aware of what my mother was doing, it could affect his position. So my mother practiced Santería, but she did so quietly so as not to anger him.

When I was fifteen, however, my father's feelings toward Santería changed. I developed a seizure disorder at that age and had seizures nearly every day. The disorder became so severe that I could not stand without having a seizure. Because of my father's position in the government, we were able to afford medical care, so my parents took me to the hospital in Santiago.

The doctors did so many tests on me that both of my arms were bruised from the needles. They also connected wires to my head, but they couldn't figure out what was wrong with me. Next, they gave me medicines that made me tired and sick. I felt worse, and my seizures still hadn't stopped. I went back home, crying all the way. When we got home, I went to my room and cried myself to sleep.

The next day was a Saturday. My father was home and got up early and sat by my bed. I remember him sitting next to me when I woke up. When my mother came into the room, they began making plans to take me to Havana, where there were better hospitals and doctors. The trip to Havana would take fourteen hours by train and would cost more money than my family made in a month. My mother was going to make the trip with me while my father stayed home to work. When I heard them making plans, I started crying again. My mother looked at me with a sadness I'd never before seen in her face, nor seen again since that moment. She looked at my father and told him to leave the room. She closed the door, put her head in my lap, and went into transition. She began singing again in that different voice and language, and again her facial expression changed. She prayed and sang and prayed and sang, over and over again. That went on for several hours. I never had another seizure after that day.

My father's feelings toward Santería changed that day. He never became a devotee or had a consultation, but he treated my mother differently and never gave her another argument about keeping pictures of San Lazaro in the house.

I was forty when I went into transition for the first time as an adult. I'd had transitions as a child when my grandmother and other devotees and santeras celebrated *bembé*, but I have no memories of those times. The first time I can remember was when my wife's sister Irma was gravely ill in the hospital. The official diagnosis was H1N3 flu. Irma had been in the ICU for over a month, hooked up to a machine that was breathing for her through a tube in her throat. Her fingers and toes were an ugly purplish-blue.

My wife and I were sitting in the ICU waiting room when I noticed a man staring at me. He was medium in height, with light brown skin and broad shoulders. He was also a little fat. I asked my wife, "Why does this guy keep staring at me?" I thought maybe he was gay. Then he walked over and told me his name was Carlos and said, "You have the gift. I can sense it. What you have is big, and you should get to know it. We are brothers."

When I looked up and said, "We are not brothers," Carlos responded, "We are brothers in the faith." I was sitting there with my hands palms-up on the arms of the chair, and when I looked down, I saw blue flames coming from both of my hands. That had never happened before. I instantly knew what he was talking about. Then he told me that he was a santero and was at the hospital because one of his family members was sick. He asked me to come to his home, only a short walk from the hospital. I agreed, and we walked to his home.

I have never met a santero more powerful than Carlos. His house was spectacular, decorated everywhere with beautiful tapestries and paintings of saints. Carlos is an amazing artist whose work has sold all over the world. When he took me into the room where his altar sat, I could feel great power there. It was almost overwhelming, and unlike anything I had ever felt with other santeros. Over the years, I've had consultations with other santeros, and I could feel the presence of the saint's power when I entered the room where their altars were located. With Carlos, it was stronger than anything I had ever felt before. He sat down on the floor in front of me and asked me to sit down, too. He crossed his legs in front of him in a way you would not expect a man his size would be able to do. Then he lowered his head and immediately went into transition and started singing in the same language my mother had sung to me as a child. When I heard his voice, I immediately went into transition myself. (Carlos has a beautiful voice. I hope one day the world will hear it. It's calming and stimulating at the same time and like no other music I've ever heard.)

He sang for some time, and then he made a small white doll. We both left transition, then left his house and went back to the hospital, where he explained to my wife's parents and her sister's husband that he was a santero. He asked if he could try to help my sister-in-law. The doctors had tried everything they could: the machine had been breathing for her for more than a month, but there was no sign that she was going to improve. My mother-in-law, who knew many santeras, had not asked a santera for help. They agreed to let Carlos try to help.

When we told the doctors we were going to try to help with Santería, they hung curtains around my sister-in-law's bed to separate her from the other ten patients in the room, and then the doctors and nurses left the ICU. Santería is common in Cuba, where an estimated 95 percent of the people have some level of belief and experience with the religion. Many Cuban doctors have personally witnessed the power of Santería and often have their patients consult with santeros before coming to the hospital. Once patients are in the hospital, the doctors will usually allow santeros to work without interruption.

Carlos and I went to my sister-in-law's bed, and he bowed his head and went into transition again. To this day, after meeting many santeros and santeras, I've never seen anything like what Carlos does. He can go into transition at a moment's notice and control it completely. He began singing again, and I went into transition again.

He brought out the white doll he had made and began moving it around my sister-in-law's head, arms, and chest. As he did this, I began feeling sicker and sicker, and when he passed the doll completely over her chest, I collapsed to the floor. I struggled to get back up on my feet and began breathing harder and harder. I had great difficulty catching my breath. Carlos didn't stop singing and moving the doll. He continued passing it over Irma's abdomen, and when he reached her legs, I could no longer stand. But Carlos didn't stop. He passed the doll over her feet, which made me lose consciousness. I don't remember anything after that.

I learned later that when I became unconscious, Carlos ran out to the waiting room, where my wife, her parents, and Irma's husband were waiting. They could see in Carlos's eyes that something was very wrong. They called for several men and nurses and carried me outside. They put me in a car, my wife got in, and Carlos drove to his house. He and my wife carried me into his home and into the room with his altar, where they laid me face-up with my head in my wife's lap. Carlos sat down, bowed his head, and went into transition, then began singing again. He sang for over four hours, during which time I gradually improved. I began sniffing my wife's chest and pointed at her, and, then I told Carlos my sister-in-law was going to die.

After four hours at Carlos's home, I felt nearly normal but exhausted. We all went back to the hospital, where we found my in-laws still waiting, but now my mother-in-law was crying and my sister-in-law's husband was sitting with his head in his hands. My sister-in-law had died while we were in Carlos's house. Her lungs had collapsed. The doctors could not save her. My in-laws held no resentment against Carlos. They knew we had tried everything we could, and the doctors had already said there was nothing more that could be done.

I asked Carlos why he had used me to try to help my sister-in-law. Why couldn't he have done the ritual alone? He explained that when the patient is unconscious, the healer will often use a family member or someone close to the person to serve as an intermediary. This is sometimes done when the person needing help is far away or to help people when they need to make an entire life change. Sometimes animals are used as substitutes, usually a black or mixed-color chicken or goat. The chicken must be one whose feathers cover its entire body. The animal is slaughtered, and the santero will drink the animal's blood, and its heart will be removed. The santero will usually come out of transition shortly after this ritual, and everything will be normal.

After that experience with Carlos, I became much stronger in my gifts. I can now go into transition much more easily, but I still cannot exit transition without the help of a santero. Though I never became a santero myself, as a devotee I have discovered gifts I've had from childhood. I often have good and bad intuitions, intense clairvoyant moments that occur after what feels like lightning shooting through my body. Usually, these things happen around events that are important in the lives of people I interact with.

One day, my wife had three visitors at our house. The three women had recently completed their nursing studies and had taken the final test to permit them to work as nurses. They would have to wait one week before the results of their tests came back. I felt a good intuition for two of the nurses and a bad intuition for the other one. I told the first two, "You two have passed," and then turned to the other and said, "You have failed." This woman laughed at me. A week later, when the results of their tests came back, the two I'd told would pass had passed, and the one that I had told would fail had failed. She told me, "Don't ever tell me anything again. We had already slaughtered a pig to celebrate." She was never able to pass the test.

Another time, I was riding my motorcycle down the road and saw a friend talking to another man. I stopped to talk to my friend, but when I got close, I received a bad intuition. The man talking to my friend was mourning the death of another friend, who had died the day before in an accident. "Do not be so sad," I told this man. "You are going to die soon." Two days later, I saw my friend again, and he told me the other man had died in an accident the day before.

I don't like giving people bad news, but it's not something I can keep inside. When the intuitions come, I have to speak. When my daughter was studying to be a doctor, she needed to pass an important test to be able to start medical school. She had been studying for months, but she was struggling with math. About a month before the test, I told her to stop studying because she was not going to pass the test. She felt sad, but she knew I had intuitions and that they always came true. Despite what I told her, she kept studying and worked as hard as she could, but sure enough, when the results came back, she had failed.

Sometimes my intuitions are bad, even with my family. That's one of the difficult things about having gifts from the saints. Sometimes you don't have a normal life. You know things, both good and bad, about the people you love and care about. You often know in advance both the accomplishments and difficult times that normal people can experience as they come. This knowledge can make coping with challenges difficult. Because I may have a bad intuition and know something bad is coming, this puts me in a bad mood, and I have difficulty explaining to my wife what is wrong when she can sense the change in my mood. It's something you have to get used to. But in the end, I think it's a burden worth having and that it's better to have the gift than not.

Whenever I meet a santero, they always ask me why I do not become a santero, too. Why do I remain only a devotee? They often tell me they can sense the power in me, that I'm stronger than some santeros. I have thought about it, but I have decided that being a devotee is the right choice for me. Being a santero is a sacred trust. It is a life of service, and I know myself. When someone has that much power, I imagine there is always the temptation to use it for selfish reasons. I'd rather stay a devotee and just use my gifts to help my family.

A Taxi Driver Shares Some of His Experiences

When my wife and I visit Cuba, we like to work with the same drivers and chefs we've built a relationship with over the years as we've visited the island. One such driver who's become our go-to cab driver is a beloved gentleman we'll affectionately refer to as C here. Never the one to be too shy to strike up friendly banter, I had to know the story of a small doll resembling Aunt Jemima I observed C* had hanging from his rearview mirror. So began what ended up being the first of many interviews with observers or non-practitioners both within and outside of Cuba. As a person in the tourism/service industry, C* had many stories he was eagerly willing to share. Below are just a few of his personal accounts.*

I've seen a lot of changes in Cuba over the last twenty years. Things are much better now than when I was a young man. When the tourism industry began flourishing about ten years ago, I was fortunate to start working as a taxi driver. Being a taxi driver in Cuba is a very lucrative career because drivers can make more than most other professions, including doctors. Over the years, I have had many interactions with santeras and santeros, as have the numerous passengers I've chauffeured. I will tell you about a few of them.

Shortly after I began working as a driver, my family became more prosperous. We were able to afford more clothes and buy more food than we had before I became a driver. My family previously struggled to afford things like toys for my children and new shoes. As a driver, that changed, and I was happier than I'd been in a long time.

One day while I was walking with one of my friends in a town far from where I lived, a santera who lived near this friend's house told me she needed to speak with me. When I walked over to her, she said, "You are a driver, yes?" When I said I was, she said, "You need to have a black doll in your car. It will help to keep you safe." The santera then told me that she was not able to prepare the doll, that it would have to be prepared by a babalao.

My best friend had been a santero for many years, and I'd seen him helping many people. But he was not a babalao. He made the doll for me, but he could not prepare the doll in the way I needed, so I began looking for someone to prepare the doll properly. Unfortunately, neither my friend nor any of the santeros I knew, knew of a babalao I could consult. So I hung the doll on my rearview mirror anyway and decided to give up looking for a babalao to prepare it.

Shortly after I began working as a driver, I started having many clients in Guantánamo Bay, where there are many powerful santeros and paleros. Paleros often live near cemeteries because they use dirt from the cemetery or bury things there for the rituals they do. One very powerful palero I know who lives in Guantánamo Bay frankly told me, "I'm a witch," the first time I met him and asked him what he did for a living. He lives in an enormous house. I have taken countless people there to see him over the years. Some travel from lands all around the world to learn from him and buy his services. He has also helped many Cubans who've previously attempted but been unable to successfully travel to the United States. He charges his clients thousands of dollars, and they pay it because he's very powerful. He's so powerful, in fact, he told me he has the power to kill. And I believe him. His prosperity is all the proof I need.

I frequently drive clients to see santeros and paleros. It's something I've had to get used to. At the end of the day, what people choose to do is their choice. One day I was driving a group of tourists to see a santero I'd never met before. As my passengers left the taxi, one of them asked me to wait for them. Waiting is not something I usually like to do because consultations can take hours, but they agreed to pay me for the time I would spend waiting, so I turned off the engine and reclined my seat.

As my passengers neared the santero's house, he came out and opened the door for them. He delayed going inside a moment, and then motioned to me to come in, too. "Are you coming?" he asked. When I said, "No, I'm just here to drop them off," he responded, "But you need the black doll prepared, yes?" I was shocked! How did he know? My car was parked far enough away from his house that he couldn't see the black doll, so how would he know I needed to have it prepared? "I can prepare the doll for you," he said. "Come in, and I will." I went in.

I let my passengers have their consultations first, and then the santero called me into the room where his altar was. He told me that the doll would no longer be strong enough. "You have *the bad eye* on you badly now," he said. "You will also need to boil seven bulls' eyes and then place them in a container and keep that container in your taxi at all times." I later prepared the bulls' eyes and placed the container in the trunk of my taxi. I also put the black doll back on my rearview mirror alongside a picture of the Virgen.

I drove the same car for the next seven years. I never had an accident, a flat tire, or anything wrong with the car the entire time I had it. About a year ago, I switched taxi companies and began driving a different car. The first day after I started my new job, I got a call from a friend who had taken over driving my old car. He'd had a serious accident but had survived. I asked him what happened, and he said that he didn't know. Then I asked him, "Did you see the container I left in the trunk?"

"Yeah," he said. "I thought it was just junk, so I threw it out."

In Cuba, almost every person you meet, or someone they know, has had an experience with a santero. As I was growing up, my uncle taught me to respect their power. My uncle liked to gamble. He was a hustler, though not anything illegal; he just liked to buy and trade things all the time. He always was coming up with ways to make money. Things can be hard in Cuba, and people come up with all kinds of ways to make money.

One day, my uncle went to a santero and asked him to prepare something to give him good luck gambling. The santero prepared something, but I'm not sure exactly what it was. The next week, my uncle gambled about two thousand pesos and lost. He tried again and gambled ten thousand pesos and lost. That was a huge amount of money for a Cuban back then. It still is. That same night, he got very drunk. Angry about the money he'd lost gambling, he went to the santero's house and beat him up and destroyed his altar. He left the house as quickly as he had come. A few weeks after he attacked the santero, my uncle broke out in blisters all over his body, from head to toe. He went to the hospital and was treated with medicine and was discharged a few days later, but the blisters came back within a month. My uncle went back to the hospital and was admitted again. Again, he stayed for several days, was discharged, and went home.

Two months later, while my uncle was working on his farm, a donkey kicked him and broke both his legs. He was back in the hospital again, but this time he stayed several weeks as his legs healed. Two months later, he fell off his horse and broke one of his legs again. That's when my mother said to him, "This is happening to you because you offended the saints when you destroyed the santero's altar. You need to go to the santero's house and apologize." My uncle followed her advice, and the problems stopped.

In addition to driving tourists all over the country, I'm frequently asked by local Cubans to drive them when they or their children are sick and need to go to the hospital. In one part of Cuba, the hospitals are very busy, and santeros frequently assist the doctors. There is a santero who is so good near one hospital that one of the doctors there asks the patients if they've seen him first. If the patient hasn't and is not seriously ill, this doctor will send the patient to consult with the santero first.

There was a couple who needed a ride to see their baby, who was in the hospital. The baby was very young and had developed a rash and was not feeding. The doctors asked the parents to consult with a santero because the baby's illness was viral, and there was no treatment they could recommend. Because the baby was too young to leave the hospital, the father took the baby's bottle to the santero. Some santeros can perform a consultation using objects close to a person, things that the individual uses frequently. The santero told the father the illness was mild, then prepared an herbal mixture with water and told the father to take it back to his son and have him drink it. When he got to the hospital and told the doctor what the santero had said, the doctor was so confident in what the santero said that he discharged the baby before the symptoms had gone away. The baby drank from the bottle on the way home in my taxi, and the rash promptly disappeared. The child had no further problems.

The Evolution of Religion, Indigenous Philosophy, and the Roots of String Theory

One's belief or disbelief in unseen entities, *spirits*, is in no small measure, preeminently defined by cultural influences. We are sadly, sometimes conscious of only what we've experienced it seems, and cultural influences have nearly exclusively defined the general public's perception of whether or not unseen entities are an actuality. We live in a world of dichotomy, where those from indigenous cultures and others with religious beliefs, often openly acknowledge the existence of unseen intelligent beings, while many others do not. In time, that incongruence will resolve as the actuality of the existence of these beings will become openly acknowledged because of the scientific verification of them. The central question then will become the most practical one. Since such beings exists, what role, if any, do they have in the lives of individuals and or humankind? And it is essential for humanity to know and understand such.

Appreciation of this actuality necessitates a consideration of the evolution of religious thought as such pertain to the practice and belief in various forms of spiritism, and how such have influenced the development of scientific thought. The influences of both abrahamic religious traditions and traditional indigenous religious thoughts, how these in particular have prominently shaped Western civilization and scientific thought, or been nearly entirely excluded from such respectively, is vital to consider. The reluctance to accept even considering descriptions of intelligent unseen entities, often from indigenous people historically, is a blatant representation of the obvious ignorance of racial and ethnic prejudice. The abilities shamanic type figures present within various indigenous cultures if accurate, presents a possibility for significant scientific advancement in a number of scientific fields.

A consideration of science's influence and prejudice is in itself not some moral judgement upon humanity or religion or science in particular, but rather an analytic exercise meant to establish the degree to which such patterns of thought, custom, and perceptions have influenced the belief or lack thereof as it regards unseen entities. For the religious, particularly those professing abrahamic religious views, the existence of such entities nearly exclusively takes the form of the malevolent and benevolent characterization of demons and angels, while for the nonreligious, such concepts belong more in the realm of the ridiculous and nonsensical, not a part in any serious scientific consideration. No middle ground whatsoever has existed regarding such beliefs, and as such, no serious scientific consideration of whether difficult to believe reports regarding the manifestation of paranormal experiences are rooted in truth.

Within the context of human experiences, such accounts in times long passed regarding the paranormal and supernatural nearly exclusively were documented within the framework of religious writings and accounts. Not surprisingly, the accounts of such experiences often reflect what was the customary response of the ancient world to phenomena which transcended the knowledge base of individuals. Without the lexicon or technological understanding to describe these accounts scientifically, such were ascribed to the purview of some divine being within the paradigm of spirituality. Such experiences are similar to a scene in the latest iteration of a remake of the

Star Trek movie franchise, where a humanoid species saw a starship and began worshiping it believing it to be a God. The entirety of the scientific understanding of that world society had no explanation for the advanced technology they'd experienced, and so placed such in the realm of the religious.

Something similar has occurred nearly exclusively within the modern world, where no middle ground in the perception of credulity regarding paranormal and supernatural experiences exist, where the religious ascribe such occurrences to the realm of God, and the scientist such accounts to the realm of the nonsensical. But only the deplorably ignorant persist in ignorance once confronted with the erroneousness of even deeply entrenched biases. Such is the true power of science, which through persistent, disciplined, and organized research and observation, brings order and clarity to things that heretofore were ascribed to the realm of the unexplainable. Here is where science now lies with regards to the existence of unseen entities. To approach such consideration with discipline and persistence, fueled by recurrent observations and experimentation, with a view to taking that which has been heretofore ascribed solely to the realm of the religious, occult, and ridiculous, to that where it more appropriately belongs, a reality, fueled by poorly understood scientific realities which through more thorough and persistent study and consideration, is now brought into the realm of light and clarity.

To this end, I have labored to understand the varied ways in which unseen intelligent beings represent a new frontier of scientific knowledge. In coming to understand the reality of the existence of these entities, my previous study of both religious holy books and scholarly works by ethnographic masters like Marciae Eliade and Lydia Cabrera, as well as my own difficult to scientifically explain personal experiences and observations, played an enormous role in convincing me of the unequivocal existence of these entities. Rarely in times past did history's transcendent scientific figures take time in narrative fashion to walk society through the logical progression of their discoveries and theories. For me, such seems to be as important as the discovery itself. For unlike many other breakthrough scientific discoveries, the existence of alien life in the form of unseen entities poses to hold ramifications that extend for beyond the realm of the scientific. The connection of such with religious thought and practice places such beings' significance in a cadre of importance that poses extraordinary significance for human society. In addition to the obvious scientific importance of understanding them as beings, they also represent a unique source of energy humanity has little to no understanding of.

The actuality that unseen entities are comprised of, and operate through, a type of energy humanity is completely ignorant of scientifically, is to me, the most important physical reality of these beings existence. What advances will come from further understanding of this energy, how it is produced, sustained, and can be used, holds promise that is unquantifiable for humanity. Renewable energy moves from the realm of the theoretical to the actual in a practical way, for what does it say about the theoretical possibilities that exist knowing that there are sentient beings comprised exclusively of energy? That in itself illustrates the depth to which this energy form can be organized. And if such can be organized in such a way as to sustain life, in what lesser ways can such energy be utilized, cultivated, generated, and distributed for other usage? Science is impelled by the depth of significance such holds for humanity to study unseen entities, for in unlocking the mysteries of their existence, and capabilities, an undeniable horizon of possibilities is unlocked for humanity.

While the declaration that the physical manifestations of morphologic change within the appearance of shamanic water possibly indicates these entities are distinct life-forms entirely comprised of energy, another possibility is that such morphologic change is simply the manifestation of an advanced form of technology through which these beings consciousness is projected. Many of the shamans I've interviewed indicate that they can discern the presence of these entities simply by observing their presence within this water. While such may be an indication of these entities presence, other possibilities must at least be considered.

The lexicon through which we communicate thoughts concerning unseen entities becomes an ever present matter of prime significance. It's often through the lens of analysis of other cultures and their references to these entities, that a unique picture of their nature, at least from these cultures' perspectives, provides a foundation for scientific investigation and consideration. Consider for example the word spirit, as it was used in many Native American cultures. In many Native American cultures there is or was no word for what we term energy. The word spirit was used in its stead. For example, plant life and animal life were described as having spirit, as were inanimate things like rivers and rocks. All things were considered as both having and manifesting spirit, and all things were considered as being derivatives of Great Spirit, or the divine being from which all things originate. This spirit, or energy, present in all things, also binds all things together, and we see elements of what more appropriately might be considered as universal string theory, long before such came into the purview of theoretical physics. A sober realization of the obvious value of such elevated cosmological perspectives of these and other indigenous peoples, elevates groups who have customarily, flippantly been ascribed as ignorant people, into a perspective much more in harmony to what these peoples actually are. The indigenous of our world are the possessors of unique forms of cosmology and technology capable of affecting the physical world in which they are a part through their understanding of energy and unseen intelligent beings.

A Santero's Red String

*AC**

I was around ten or eleven when I noticed that things that were not normal were happening to me. I would sense when spirits were close to me. They would talk to me and put their hands on me when I was lying down.

How did you start working with your gift?

Thirty-two years ago, I had chronic hepatic cirrhosis with nine liver nodules. A doctor told me I had about five years of life left. With direction from the saints, I diagnosed the liver problem myself and started treating it. The spirits would tell me exactly what I had to do, and I would do it. That's the reason I am still alive today. I am now sixty-two years old, and it has been thirty-two years since I was diagnosed with cirrhosis. When I started treating myself, I first removed the inflammation I had in my liver with the spirits' directions. In time, my liver went down to normal size, and the doctors were amazed. They couldn't explain how that happened. From there, I started helping my family.

Over time, the spirits have given me secret prayers I cannot share. They are for me and must remain my secret until I die. For example, the way I heal people from a distance is through a secret prayer. I can transmit my healing to an individual anywhere in the entire world because of this prayer.

From here in Cuba, from my own land, I can feel if or when a person feels better, and if I have fixed their medical problem. When the illness leaves, I can feel it through a red ribbon I use. The spirits have guided me to heal with that ribbon. When I see a person face to face, I can discern through measurements taken with the ribbon of the person's arms and other places exactly what problems they have.

The same thing happens when I'm consulting a couple with pregnancy problems. With the red ribbon, I can detect pregnancies even before a doctor can see the pregnancy with ultrasound at times. Many pregnant women have come to me before they knew they were pregnant, and sometimes even after a doctor told them they weren't. The pregnancy was simply too early to detect with medical devices. A lot of babies have escaped death because of the guidance curanderos gave to mothers.

Every day, I see forty to fifty people for various problems. It's busier than a medical clinic. They come from all over Cuba to see me. A nurse who came to see me told me that I work harder than the hospital doctors do.

You said when you were young, you felt like something would touch you?

I felt like something was pulling me toward it. I could sense emotions in people and other things. For example, when I took tests in school, I would think it's one answer, and then the spirits would tell me it's another answer. I would do what the spirit was telling me, and it would always be the right answer. Another time, I got sick for fifteen days and went to school just to take the tests I'd missed. I passed the tests, and the teacher set me in front of the class as an example of what could be achieved. She thought I was studying, but the spirits were just giving me the right answers. I knew it was abnormal, and I didn't know what to do with it, or why I could do it.

Is there an ethical standard you hold yourself to?

In this type of healing that we practice, the essential thing you *cannot* do is make things up or lie. It is very dangerous for us as curanderos to do that. The saints would punish me if I were to lie or mislead someone. Whatever problem a person has, it is my responsibility to tell them. I also believe a healer should not brag about themselves or about the things they've done. It is only with the help of the spirits anything is accomplished.

Some people like to work with spirits to do bad things, and others to do good. I don't like anything bad. I trust in God and the Holy Spirit a lot. The first thing I do is ask for permission from God and the Holy Spirit before I do anything spiritual. To be able to go into transition, I must always pray before.

One of the positive things I do is that I do not say no to anyone. I serve everyone. I have never sent anyone away. I am here to serve, and I do not charge for what I do.

What do you feel when you're healing someone?

I feel the energy currents of the saints when I heal. It's a beautiful feeling, and I feel very spiritual. When I heal people from a distance, the currents feel different. They feel long and big. It's very invigorating to feel that energy when you do that work. When I heal someone or remove someone's illness, I feel this contentment from knowing I have completed what I was meant to do. I thank God for giving me this gift and the knowledge to complete the work I do.

What do you consider yourself to be?

I do not consider myself to be anything. I do my healing work because it's something that comes with my spirituality, and it's for the good of humanity. I consider myself a normal person, just like the rest of us, only with a gift and knowledge that God has given me to help my neighbor.

In your family, was anyone else involved with healing or working with the spirits?

My mom and dad were devotees of their own saints. My parents would pray to their saints and light their candles, but they never went beyond that. I have a lot of saints. I have Santa Barbara, San Rafael the divine doctor, Santa Rita, and San Lazaro. I have a little temple where I worship. I ask them for power. There are more than thirty saints on my altar.

I don't think of myself as better than anyone else. I respect all religions. All religions are good, but most people are not sincerely religious. That's where the problem lies.

Which is your foremost saint?

The Virgin Mary, the Mother of God. I have her as Patrón of my altar.

From the age of ten to twenty-eight, which spirits were bothering you?

I was about twenty-eight when I realized my gift was developing. Before then, I felt like the saints were continually guiding me. They were leading me, teaching me, working on me trusting in them, and on them relying on me. They were preparing me for what I was going to become.

When I started with my family, I started with my mom. She had terrible varicose veins that would rupture and bleed everywhere. I would go to the hospital, put my hands over the bleeding wound, and it would close and stop bleeding. Even after I did that, we would still take her to the hospital, but the doctors told me to stop bringing her to the hospital because what I did was all she needed.

Next was the time my dad was in a coma. He was so sick they were going to send him home to die. He lived in Havana at the time. I went to Havana and told my brother to let me be by myself with our dad. I had a strong intuition telling me that I would cure him, that I was going to make him better. Everyone left the room, and I was there with him. I waited till the other patients in the hospice were asleep, and I stayed awake next to him after I got a special license from the head of the hospital to stay with my dad after visiting hours.

I said a few prayers, and my hand felt like it had become ten times heavier than usual. I put it on my dad's chest. I sat down in a rocking chair next to his bed after I was done. Around 2 a.m., I felt my dad wake up. He looked around the room and asked me, "Excuse me, sir, do you know where the bathroom is?"

"Dad," I said, "you don't recognize me? Let's go," I said. "I'm going to take you to the bathroom."

Nurses and doctors came running, asking each other, how could this happen? I took my dad to the bathroom, and he pooped so largely, the whole hospital smelled. His poop was as white as sand. The doctor discharged him the next day. He was almost eighty years old at the time. He lasted fourteen more years after that.

So, what disease did he have?

He was admitted with Hepatitis A. Because of his age, he had a very bad case.

Another of my early cases was my cousin. They were going to take her to another hospital because she had severe problems with her heart. She had two heart attacks. From far away, I saw her in the hospital. I said my prayers, and she reacted, and now she is good—no more heart attacks or heart problems. That also helped me realize what I could accomplish from a distance, not just when someone is near me.

People would call me from Havana, and from my house, I would do my work and heal family members. For example, they would call me if my brother was sick. I would do my work, and then they would call me back and tell me he was better.

More and more, I began noticing what I was capable of doing even from far away. There was a kid in Holguin whose dad called me around midnight. A doctor had given him my number to remove whatever the kid had. The boy was in the ICU, unconscious. He was seriously ill. The father wanted to pick me up in a car and drive me to the hospital, but I knew it would have cost him a lot of money. So I told him, "Let's do something. Give me the name of the child, the first and last name, and date of birth. From my home, I will try to heal him." I went to my temple and did what I know how to do. After thirty minutes, I called the ICU and asked the nurse about the patient in bed #1. The nurse told me, "The boy is now awake. He has opened his eyes. He's better." After fifteen minutes, I called again, and the nurse told me, "The boy is drinking juice. His mom is sitting next to him." I called the father and told him what was happening with the boy. The father couldn't believe it. He was in his car, ready to pick me up. I told him to go instead to the ICU. He called me, crying in gratitude for what I had done.

So you work through your altar, right?

Yes. It's not big, but it's full of saints. That's my energy, the source of my power. I base everything on the Holy Spirit, but they are the ones that help me directly. They are my strength.

When did you start making your altar? When did you know you needed one?

I don't remember exactly how old I was. Over time, I felt like there were a lot of things I needed. I needed San Lazaro to help people with skin problems. I felt like I needed him, so I got a little San Lazaro. And, of course, I went to buy my Virgin Mary.

The more good I do, the more strength I get, and the more God helps me. Everything I do is accomplished by the spirits guiding me on how to do it. There are a lot of santeros who are not real. Like the Bible says, false prophets that make a living off this work. I don't think of myself as being like God. I feel like I am helped by God and sent by God. I just want to help everyone that needs me.

Some santeros were not born with it. You go, and they can destroy and harm others. Only the bad are the ones who harm. I don't like that.

Do you have the ability to see things in a glass of water?

Yes, I can see in a cup of water. Having vision is another gift. I have a wine glass I use when I work with water. I invoke the being I want to work with, and I can see everything right there, reflected in the water. For example, there was a girl who came for a consult. I saw a motorcycle in a cup of water, meaning she had come on a motor-cycle. I noticed she had a black cross wrapped in a black piece of cloth inside her purse. I asked her about that. "What is the black cross?" I asked. She told me, "Don't worry, it isn't Voodoo."

A lot of santeros are known because they can see the future. There's a lot of times when a spirit tells us what's going to happen. Sometimes I throw the shells before a person comes to the house. I don't like to do that often because everything shows up with that. Deaths of family members, for example, that the person didn't ask to know about. I also throw the coconut to see the future.

Do you take people to be godchildren?

Yes. I have a few godchildren in Cuba and the United States. I accept someone when my altar chooses a person, or the spirits discern that they have earned the right. Sometimes people ask me, and I say no. Sometimes the altar says no to a person.

What is your relationship with, or perspective on, religion? Do you have a religion?

I like the Christian religion. I consider myself a Catholic.

There's a lot of people who have gotten sick when they left their saints. Has anything like that happened to you?

My saints are my strength. I worship them a lot. That has never been my experience because I have never left them. I have grown to have deep respect and love for them. First of all, they saved me from death. They put me on Earth. I would have been dead thirty years ago, and I wouldn't be telling you this story without my saints. I'm still here serving humanity and doing what's right. My life itself is a gift from my saints.

How do you celebrate bembé?

The first thing we do is ask for permission from the saint before we do anything. We have to ask if the saint accepts the celebration. Then we do a cleansing on each of the members who are present for the ritual. Cigar smoke and branches of leaves of certain trees are passed over the person. Usually, the godchildren of the curandero or santero conducting the *bembé* are cleansed first, then everyone else. We do the cleansing to remove any bad energy from them and then ask the saints for health.

A lot of the people present at the *bembé* make promises in front of the saint. Most of them make promises to San Lazaro for health, to be healed from illness. Many promises are made with regard to having a *bembé* the following year. For example, a person will promise that they will hold a *bembé* for the saint the following year if they are healed.

My *bembés* are always for San Lazaro, Babalao Aye. There are different San Lazaros, so it's essential to be specific. After the prayers and cleansings, we start playing the drums.

Have you been in transition?

Yes, many, many times. On many occasions, when things are really bad for the person I'm consulting with, I have to bring the saint down and work with the spirit. You have to do what the spirit says. The person has to have a pencil and a notebook to write down everything that is told to them when I'm in transition, because I cannot remember it. When the saint leaves my body, my mind is clear, and I feel clean emotionally. Once a spirit has left you, you also see everything physical with clarity. It doesn't matter if it's cloudy outside or raining. Everything is bright and clear and beautiful.

Can you explain what happens when you're getting into transition?

From the beginning, when the saint begins to enter my body, I start feeling the presence of the spirit. I can sense a current moving inside me and feel the saint's presence. Then I start having physical changes. My tongue becomes stiff, and I start talking differently. Then I just go numb and don't remember anything until the saint leaves my body. I have seen many people who claim to be in transition. Sometimes they will put on a show at someone's *bembé* or act like they're in transition to try to swindle people. I'm a curandero, and I know when someone is truly in transition. When I see the ones faking it, I tell them, "Just stop. Get up; there's nothing in you." *Bembé* and the rituals we do are all about working in harmony with the saints. They are sacred beings, not something to be faked.

How can you tell when someone is genuinely in transition or possessed by a spirit?

The way you know, you have to look in the cup of water, and you can see the spirit in the water. The water becomes turbulent. Other people standing nearby might see it as just clear water, but the ones who have the vision can see the spirit's presence in the center of the water. A lot of people don't have a vision for that. When the spirits come down before they enter a person, the spirit always approaches on his knees. The spirits don't approach standing.

At my *bembés,* I ask the saint for no one to go into transition. I also do a ritual so that it will not happen. I prefer everything to be calm and peaceful, to go about my work of healing and helping the people without distraction.

After the experiences you had when you were young seeing spirits and feeling their presence, how did you develop your gift?

Things happened little by little. I didn't know what I had. I would get heart palpitations, goosebumps, and just weird things would happen as I would sense the spirits' presence around me. My mom took me to a santera. They called her *the Dead and Alive.* She told my mother, "Your little one has a strong gift. Do not let anyone touch his head. Keep him away from the paleros."

When I was young, I was in Guantánamo, visiting my aunt, and she took me to see another santera. She told me the same thing: I had a powerful gift, and I would help many people in time. Little by little, I started developing my abilities, and the spirits would talk freely to me. I could feel the presence of a spirit next to me all the time.

For my work, health has always been the priority for me. That is how I was able to heal my mom, my brother, my sister, and my father. I had a niece who couldn't conceive with her husband. She now has two kids. Many US residents have come to me with fertility issues, and now they have several kids.

When I was in Santiago, I was next to a santero who could feel that I had power. We decided to travel together, and while we were by the side of the road, I told him the exact car that would pick us up while it was still far away. A spirit told me the driver would be going where we were headed and would help us. It happened just as I told him. We became friends after that because he also works only for the good.

Do you also feel the power of other people?

A guy from the United States came to see if I was a real curandero. He had heard about me, and he thought he was stronger than me and that I was just a fake. He came to my home. He said, "I need a consult." I told him, "I can't. I have other things to do." I already knew his intentions. He told me, "No, no, no, I need you to see me now." I leaned toward him, put something near his hand, and hit him in the chest. He started crawling on the floor, out of his mind. Then I removed the spirit he had inside, and he came back to himself and left.

Some people need that. I felt like he was disrespectful. He came back the next day. I told him his problems, the things happening with him, and why he was having trouble in the United States. He has his gift, too, and we have since become friends.

He asked me if I charge. I told him I do not. Health doesn't have a price. He put \$20 down for my saint, gave me a large cash gift, and said, in the US, this is how much what I did for him is worth. "You have told me everything that no one else has told me," he said. "I thank you for what you have done."

Why do I feel how I am feeling now? I feel a presence from my neck to the top of my head. Why do I feel that when I'm talking to you? I feel, like, energy currents and something heavy on me.

Are you working with something?

No.

You're asking about things I shouldn't talk about. A lot of things are sacred. There are certain things that spirits are telling me, and you need to write only the basics. Do not ask for more details. Some things must remain hidden.

An Atheist Learns to Grieve
AC*

The "ring of truth" is an interesting expression. It speaks of a conveyed feeling of truth in a person's eyewitness account or experience. The account below is heavily laden with the ring of truth, and there are various aspects of the account that were observed by multiple eyewitnesses. I challenge the reader to consider the various rational explanations for what this eyewitness describes.

I've been an atheist my whole life. After the death of my daughter, my wife started reading the Bible and going to church. Religion was never something that appealed to me, but I began reading the Bible to understand what my wife was learning and to be able to better empathize with what it was she felt when she went to church. She took the death of our daughter much harder than I did. I grieved and was eventually able to move on and find a measure of contentment in my life. I had to be strong for both of us, and also for our other daughter and newly motherless granddaughter.

I began having severe chest pain a few years after my daughter's death. At times, the pain was so severe I couldn't walk. I knew something was seriously wrong, but I didn't want to worry my wife, so I didn't talk about it, and I didn't go to the hospital.

One day when I was walking down the street, a santero well known in the province saw me. He walked over to me and said, "You have a very serious problem with your heart. You are having chest pain often, yes?" I was shocked, but I admitted that he was correct. "Walk with me to my home," he said. "I will help you." Skeptical and reluctant, I followed him to his home, but I had heard he was an excellent santero, and that set me at ease some. I entered the room with his altar, unsure about what he would tell me. He went into transition and said to me, "You need to go home and cry."

There were no special potions, no singing, no statements of anything significantly special. And yet, when he spoke those words, I knew he was right. I went home and gave way to tears for the first time in my adult life. I felt a release I didn't even know I needed. I had been so focused on staying strong for my family, I'd had no time to fully grieve. My chest pain resolved that night, and I haven't had a problem with it since.

I'm not a religious man, but I know the spirits are real, and santeros have power. Sometime after my first experience with a santero, a longtime friend of mine died. He was a renowned neurologist in the community and had many friends in the city. It's common in Cuba to invite a santero or santera to the house on the day of the funeral. They often serve as a medium during gatherings held after the funeral. Usually, the santero will channel the spirit of the person who has died and will talk, move, and act with the characteristics of that person. People at those gatherings are often lightened and encouraged as they interact with the santero channeling the spirit of their dead loved one. This is a common occurrence in Cuba.

On this particular occasion, I went to the home of my dead friend, and many people were gathered there. There was also a considerable number of children present, and when the santero entered the home, they were sitting and playing in the living room. When the santero entered an adjacent room, the children began singing loudly, dancing, and jumping in a way that frightened me. It was unusual for me to see children moving in this way. It was very disturbing. I had been to other funeral events but had never seen anything like this. Then the santero made his way toward the living room and stood in the doorway between the living room and another larger room. The children immediately calmed down and stopped singing. He called them one by one, and they walked quietly into the larger room. The children sat in the middle of the floor in the larger room and began calmly talking to each other. I still felt uncomfortable watching something like that happening. I wanted to run out of the house. To this day, I've never seen anything like it.

I was with my neighbor at the time, and after the santero walked the children back into the living room, we went into the kitchen to get some refreshments. I still felt uneasy when a voice whispered in my ear, "You haven't seen anything yet." I turned to look to see who was talking to me. It was a person who looked exactly like my friend, the neurologist whose funeral we had just attended. It was not the santero. And again, I wanted to run out of the house. I looked at my neighbor, who was more accustomed to the experience than I was and must have noticed my fear. He told me, "We're staying until midnight." For the next several hours, I watched the santero channel the spirit I'd seen, someone who looked exactly like my recently dead friend. The santero walked, talked, and interacted with people the same way my dead friend would have done if he'd been alive.

It can be a difficult thing for someone from another culture to appreciate the truth and the value of this. I was scared beyond belief during my first experience at a funeral like this. Over time, however, you get used to it and accept what santeros can do. Viewed from a different perspective, we can more easily appreciate why such a ceremony helps to bring relief to the grieving family and friends. Death is oftentimes unexpected, but even when it isn't, there's a finality to it that punctuates the loss for those who've lost someone. It hurts in such a way that the loss we feel from death is painful in a way no other loss is. Being able to have a little more time with the essence of the person, how they talk, move, act, and think, even if we know it's only temporary, can provide relief and comfort in a way a traditional funeral without a santero cannot.

The Bad Eye and the Power of Faith

*AC**

I've been a physician for over thirty years. After finishing eight years of medical training, I started working in a local hospital as a specialist in internal medicine. Cuba's medical system is different from other countries, particularly the United States. Doctors in internal medicine not only take care of patients in the office or hospital, but every week, I'm also on call to take care of patients in the emergency room.

I had one of the most challenging cases of my career shortly after I finished my medical training. There was a young girl, six years old, who came in with an acute diarrheal illness. She had a high fever with diarrhea and vomiting and was severely dehydrated. I knew she needed to be transferred to a larger hospital with a pediatric ICU and equipment appropriate for a critically ill pediatric patient.

After consulting with a team of physicians in the hospital, we concluded that the girl had cholera. Though not common in Cuba, in the summer, when it's dry, it can happen. We began treating her. The most critical treatment was fluid resuscitation, as the girl was obviously dehydrated and also appeared septic. The transport team was approximately an hour and a half away in another city. It would take them nearly four hours to complete the transfer. By the time they could arrive and pick up the child, and then transport her to the larger hospital, precious time would pass. With a critical and unstable patient, that was a less-than-ideal choice.

The small hospital where I was working that night had an extremely limited supply of vasopressor medications. When the young girl didn't respond to fluid resuscitation, and her pressure remained critically low, both the medical staff and the family became more and more concerned. The family insisted that they be allowed to bring a santera to help care for the child.

The santera they knew was named Toña, *the Living Dead*. She was a popular santera in the community. She'd had a heart attack in her fifties, and afterward, her heart flat-lined for several minutes. She was pronounced dead prematurely. A day later, she woke up during the wake that was being held for her and afterward became a santera. It took Toña about an hour to get to the hospital. By the time she arrived, the child had worsened. The transport team was on its way, but still some distance from the hospital, and I was worried the child would die in transit without the necessary vasopressor medication.

When I first saw Toña, she was walking slowly up the hill that leads to the hospital from the coast. She was smoking a cigar, and when she entered the hospital, the aroma filled the waiting room. The family went to greet her and then brought her to the child. I was hesitant to let her do anything to the child, but when she came closer to me, I could feel something unique about her presence. She looked me in the eye and, with confidence, said, "Now you will see the power of faith do what your medications cannot."

I disregarded all the policies I knew that the hospital would want to be followed and asked her what she wanted me to do. The child was dying, and there was nothing further I could do. I took her to the child, who was now in a small room at the end of the emergency department. The family and several of my physician colleagues were in the room. I told Toña she could begin, but we had one restriction: the child could not be given anything by mouth. Toña took a look at the child, sat down, and laid the girl's head in her lap. She then took out some herbs and began praying and rubbing the child's head with the leaves. She said the child's illness was the result of *the bad eye*. She asked that the child's IV be removed, and the other equipment attached to her, a pulse-ox monitor and blood pressure cuff, also be removed.

I looked at the nurses and other physicians and reluctantly agreed. At this point, I didn't see what good they were doing, anyway. Toña began praying and singing and blowing cigar smoke on the little girl's face. She did this for the next approximately thirty to forty-five minutes as the family and the medical staff stood back, watching her work. The other doctors and I talked and waited, observing carefully. As we spoke, we learned each of us had had our own experiences with santeras when we were children.

As Toña sang and prayed, the child's color and appearance gradually and steadily improved. By the time she finished working, the child was awake, and her temperature and other vital signs had returned to normal. I watched the once flaccid, unconscious child stand and walk around the room as if nothing had been wrong with her just a short time before. By the time the transport team arrived, the child was ready to be discharged. I was grateful for what the santera had done. The family surrounded Toña with smiles and expressions of gratitude, and she left the hospital as unassumingly as she had come. But before leaving, she thanked *the Virgin* in front of everyone for what she was able to do through her.

I don't know how Toña was able to heal the child. It makes no sense medically. Yet with no medical training and no degree, she was able to do what I could not. When I spoke with Toña later, she told me that many children have died as a result of having *the bad eye*. I have thought to myself many times since that day, how many children could be saved if science and santeros actually worked together?

A Nurse Experiences the Unexplainable

*AC**

I was raised in a home where Santería was practiced openly. As a child, it was normal for me to see my family and neighbors during *bembé* and my grandmother slaughtering chickens, roosters, and goats. It was routine for me to watch the people who loved me participating in various Santería ceremonies. My grandmother was over 100 years old when I was young, but she would still participate in *bembé*. She took no medicine and didn't wear glasses. Many generations of the women in my family, including my grandmother and mother, were santeras.

My mother had a simple altar. It was made of wood and had two saints on it, San Lazaro and the Virgen de la Caridad. Around the altar were white flowers, and my mother put a small bowl with several cigars and *turron de coco* (a sweet made from sugar, coconut, and lime juice) and placed it in the middle of the altar. Every day I watched her take care of her saints. She would wake up early, light a small white candle, and place it on her altar and pray. She would do that before she made breakfast or did anything else. At times, my father would steal the sweets from her altar, and my mother would yell at him: "Your mouth is going to turn upside down!" He was an atheist and didn't believe in Santería, but nothing ever happened to him. For Santería to work, I think you have to have a measure of belief in your mind. Plus, I'm sure my mother asked for forgiveness for him many times.

I know Santería has power, and I know most santeros are good people, and the good santeros do is real. But I don't believe their power is from God. I see starving children in the world and all sorts of problems everywhere. I don't think God would concern himself with telling people about simple things in their life, like if their wife or husband is cheating or whether they should change jobs. The people helped by santeros eventually die anyway, no matter what santeros do. People can become obsessed with Santería, and some think it can solve all their problems. I've seen this many times. Santería is not a cure-all; the saints have their limitations. Only God can genuinely fix everything, and saints can't always answer fundamental philosophical questions.

I started studying the Bible at age fifteen with Jehovah's Witnesses, and the more I read, the more I realized I should pray only to God. The Bible teaches us about life, how to be a good person and raise our children. Being a Christian is not like believing in Voodoo or Santería; it's a way of life. It teaches you how to live and treat people the right way. Even though God doesn't speak to me the way people say spirits speak to santeros, I feel close to God. When I pray and read the Bible, I feel better about life and my anxieties and problems. It also teaches me about the future, the truth about the dead, and the hope for everlasting life in paradise on Earth.

I watched my mother and grandmother help many people, but I believe that the things they accomplished came through the power of belief, from the power of the mind, not from any special relationship with spirits.

When I was young, my mother would do rituals and tell me over and over again that nothing would ever happen to me because of all the protections she had done for me. Many santeros I've met through the years have said the same thing about me: that there is powerful protection around me. I have always felt safe in my life. Even when problems are happening at work, or difficulties in my family life, I have always felt like everything will be okay.

I know that santeros and paleros can help people because I've seen things happen in my life I cannot explain logically. I'm a nurse, and I often take care of very sick patients. But my experiences with santeros have come from my own sickness and difficulties.

When I had my daughter, I had a complicated pregnancy. I hemorrhaged during the delivery, and my hemoglobin dropped to 3, about one-fourth the level of what it should be. At this level, I should have felt very ill, but I didn't. According to my family, I walked around the hospital with my shoulders hunched forward. They kept saying I was walking like San Lazaro.

While I was ill after the delivery, my family gathered in the hospital. Because the blood supply was limited, I did not receive transfusions of large amounts of blood. I got tired of being stuck at the hospital and left a few days after the delivery against my doctor's advice, despite my hemoglobin being so low. My grandmother prepared an herbal drink for me to take. It was made from condensed milk and herbs, and within a few days, my hemoglobin went up twelve points.

The doctors couldn't believe the results of my bloodwork and told me that it should take months for my hemoglobin to rise that much. They made me retake the blood tests several times, but the results were always the same. Without any logical explanation, my hemoglobin was normal.

Sometimes what santeros do isn't just physical, and that wasn't the only time I saw something happen I couldn't explain. Five years before I left for the United States, my mother told me that I was going to go. She told me not to tell anyone and to start saving money and making preparations. I didn't believe her at first. Traveling wasn't something I'd thought about or even thought I wanted at the time. But I had seen her do many things for people, and I knew that if she said it, it was going to happen. She told me to apply for a visa and that in time, the application would be accepted, and I would be allowed to leave. Five years later, I boarded a flight to Miami and moved to the United States and have since become a US citizen.

I believe in the power of the human mind. There are many things we don't understand about the mind and what humans can do. Santeros have a vast knowledge of natural medicine. I know that what santeros do is real, but I believe it's through the power of belief and natural medicine they can do these things.

The Failure of Education

"I am disturbed, I am uneasy about man because we have no guarantee that when we train a man's mind, we will train his heart; no guarantee that when we increase a man's knowledge, we will increase his goodness. There is no necessary correlation between knowledge and goodness."
—*Benjamin E. Mays*

"Our greatest natural resource is the minds of our children.
—*Walt Disney*

Of all the great heroes of science who've ever influenced me, and there have been a great many both living and dead, there's never been a scientist for whom I've felt more admiration than George Washington Carver. His brilliance, matched by a magnanimous devotion to education and humanity's welfare, represents the quintessential model of noble education used for society's benefit.

Carver was born a slave. He was a sickly child, a disadvantaged beginning that spared him from the hardship of slave labor and changed the course of his life and human history. His illness gave him time to consider other pursuits, including the study of plants.

Though he couldn't read and knew nothing formally of botany as a child, his natural gifts served him well. His inherent affinity for agricultural science eventually led his neighbors to bring him their sick plants, as he quickly garnered a reputation as *the plant doctor*, nursing much of their failing vegetation back to health.

When Carver was old enough, he would journey significantly in pursuit of quality education. When he was denied acceptance to one college for merely being a negro, a term once used for African-Americans, he journeyed to another college to continue his education. He eventually earned a master's degree and became the chair of agricultural science at the Tuskegee Institute. Carver felt a kinship for poor Southern farmers, whose soil grew depleted from the constant growing of cotton, and his work on crop rotation helped them replenish their land through the use of soil-enriching crops like yams and peanuts.

Carver was most known for his work with peanuts, though he didn't consider that his most significant accomplishment. He developed over three hundred products from peanuts. Among these were peanut flour, glue, lubricants, and other diverse applications for peanut products for both personal and industrial use. But he only applied for a handful of patents. He chose instead to share his work in annual publications so that humanity could benefit from his work. Thomas Edison recognized his genius, as did Gandhi, and even the industrialist Henry Ford. Edison offered him a large salary to come to work with him. Still, Carver refused, choosing instead to remain in his humble lab in Tuskegee, pursuing his passion quietly with the commitment he'd sustained from his childhood. He died with a modest amount of wealth, but unquestionably with an awareness that he had been exceptionally benevolent toward humanity. His example lives on for the world to see, for he is one of the few truly brilliant and benevolent men in history who placed humanity's welfare above their own profit, despite possessing the brilliance and ability to pursue an entirely different course.

When we look at the world, we are in desperate need of a revolution. Not a revolution born of violence, but a revolution of the nobility of character. Our planet is sick, the fruitage of generations of humanity living without the proper wisdom and commitment needed to maintain the delicate harmony between our species and the natural world. Among our species' many failures is a failure of education, for proper education is the foundation upon which all things of enduring value are built.

My wife was born in Cuba. She understands and hates socialism and communism in a way I never could. While I always aim to understand complex things in as nuanced a manner as possible, socialism and communism have always been just political theories for me. I never had any personal experience of living under them like my wife did growing up. Her schooling was a blend of traditional instruction in reading and arithmetic combined with education with the explicit purpose of instilling within her the values of socialism, or in other words, indoctrination. The books she was allowed to read were stipulated and aggressively censored. Elected writings and poems of Jose Martí and Fidel Castro's and Che Guevara's life histories were mandated to be read and memorized. She could not progress beyond the first grade without memorizing the story of Fidel's life. When she grew old enough to understand what she had experienced as a child and the social conditions and poverty she had lived through, she understood the value of liberty and freedom in a way only a person who's lived without it ever could.

In America, we don't indoctrinate our children the way they do in Cuba. We indoctrinate children a different way. Americans indoctrinate our children by what we teach them to value, both by word and by example. It is also the absence of other essential teachings that mars American and other Westernized children's education in many ways. We focus on educating a child's mind solely, and our world displays a lack of empathy, awareness, and compassion in our children's education. And our environment and humanity have suffered dramatically because of it.

Around the United States, in virtually every public school system in the country, programs for talented and gifted students exist. Nearly without exception, these programs are offered to students who show exceptional science, math, and reading abilities. These students are pushed towards a curriculum that embraces science and technology as the most important of all subjects to be learned, and by their exclusion, students who lack proficiency in these areas are often viewed as less valuable and left to pursue other areas of study. What these curricula exclude illustrates the most significant manner in which Westernized students are deprived of the quality of education that is most essential.

In 1983, Howard Gardner, an American developmental psychologist, described nine types of intelligence:

1. Linguistic Intelligence—Not just a knack for languages, but the ability to use language in an extraordinary way. Great writers, poets, orators, comedians, rappers, and others with this gift can use language with a skill and fluency others simply cannot.
2. Existential Intelligence—Philosophers. One definition of existential intelligence is the ability to use intuition and thought to ask and answer deep questions about human existence. These individuals are often deep thinkers and desire solitude, but also feel a deep connection with others and compassion for the suffering around them.
3. Bodily-Kinesthetic Intelligence—Athletes, sculptors, surgeons, martial artists, gymnasts, dancers, and others who can move their bodies in complex and controlled ways. There is also a sense of timing and mind–body coordination that accompanies this ability.
4. Musical Intelligence—Musical intelligence manifests as the ability to understand rhythm and tone. Individuals with this type of intelligence are often good at recognizing patterns. They seek out sound and often are skilled at musical

performance and dance. Some can memorize songs quickly and play music by ear or recognize specific musical notes by ear.

5. Naturalist Intelligence—Naturalist intelligence is exemplified by people like Steve Irwin, Jane Goodall, and George Washington Carver. It's marked by the ability to recognize and distinguish between various plant and animal life and classify cloud and rock formations. These individuals frequently feel a unique connection with the natural world.

6. Interpersonal Intelligence—The charmers. We've all known them; some people are just likable. Interpersonal intelligence is often associated with politicians, social media stars, teachers, and religious leaders. They understand both nonverbal and verbal communication, are sensitive to both the moods and temperaments of others, and have a knack for knowing how to be likable.

7. Intrapersonal Intelligence—Know thyself. Individuals with intrapersonal intelligence understand themselves profoundly and can use that knowledge in planning and directing their lives. They are often deep thinkers and shy. Psychologists, religious leaders, and philosophers often manifest this intelligence.

8. Spatial Intelligence—The ability to think in three dimensions. Architects, sculptors, pilots, and people with a heightened sense of direction all exhibit this intelligence. Children who gravitate to puzzles and mazes are usually endowed with this intelligence.

9. Logical-Mathematical Intelligence—This intelligence is characterized by using inductive and deductive reasoning, formulating hypotheses, and completing complex mathematical calculations. Scientists, lawyers, mathematicians, detectives, philosophers, and others manifest this intelligence.

There is some debate whether these categorizations are actually types of intelligence or talents. To me, it doesn't matter significantly. These intelligence types are manifested in history's geniuses in various ways, often with geniuses exemplifying extraordinary ability in more than one of these areas. Our educational system is woefully inadequate at accommodating the diversity of talents and skills present within humanity. We have a system that focuses extensively on technical and external technological development, and our society reflects the emptiness of ignoring the more essential aspects of humanity's needs required to achieve harmonious sustained coexistence with our planet and each other.

When we consider a number of history's giants like Jane Goodall, Helen Keller, Albert Einstein, Pablo Picasso, Prince, Mozart, Henry Ford, George Washington Carver, Bruce Lee, Al Pacino, Josephine Baker, Maya Angelou, Bill Gates, Steve Jobs, William Shakespeare, Martin Scorsese, Abraham Lincoln, Denzel Washington, Alexander the Great, Socrates, Misty Copeland, and countless others, we see the manifestation of genius in enormously diversified ways. As a society, one of our educational system's most significant failures is that it fails to appropriately value and identify these various intelligence types and gifts in children. We often fail to provide an opportunity in an equitable way for all of humanity, and particularly, disadvantaged communities of the impoverished and people of color. If we are to achieve our optimal potential as a species, our educational system must adopt and adapt new assessment tools and flexibility to address areas in which society as a whole has been grossly deficient.

The Overlooked Intelligence Types

Empathic Intelligence
There are two significant types of intelligence Dr. Gardner failed to present in his groundbreaking work. These two deserve to be chief

among our educational interests, for their deficiency as educational priorities in our society contributes enormously to the degree of moral, ethical, and environmental toxicity that permeates our world. Throughout history, there have always been individuals who exemplified moral and ethical genius, what I term *empathic intelligence.* People like Mother Teresa, Jesus, Gandhi, Martin Luther King Jr., Malcolm X, Buddha, Mohammad, Harriet Tubman, the Apostle Paul, Jane Goodall, and others, whose genius was not only in their concern and compassion for others or for nature, but a willingness to suffer every manner of injustice and embrace personal sacrifice, even to the point of death, for the welfare and aspirations of others. There's moral courage that accompanies this empathy, and these individuals often are compelled to organize, demonstrate overtly, and sustain through persistent activism a fight for the rights and freedoms of others.

Nature is not without a sense of balance and harmony. And as the world has known its complement of psychopaths and sociopaths, individuals without bounds in their lack of empathy and concern for others, so too nature has given us individuals on the opposite end of that spectrum, uniquely and superiorly empathic individuals. Psychopaths are a blend of genetic and social-environmental factors, with a willingness to commit all manner of torture, crime, and murder; empathic geniuses, those who manifest the opposite extreme of empathic intelligence, are also likely to blend these same factors. Empathically gifted individuals are by far the harder of the two to identify, for they leave no bodies or crime sprees in the wake of their lives; instead, as is often the case, their selfless service to humanity frequently goes unnoticed beyond the circle of those touched by their benevolence.

Is it possible for our educational and health systems to develop mechanisms to identify and encourage empathic characteristics in our populace? Clearly, such an endeavor would be beneficial and worth researching and considering. As genetic analysis becomes more sophisticated, it may help us identify not only increasing numbers of physical traits and characteristics, but behavioral traits as well. DNA analysis and other studies of psychopaths may indeed help identify the genetic patterns associated with this neuropathology. As physiologic and neurohormonal systems often work through cycles and counterbalances, empathy-contributing genes and neurohormonal cycles might well be identified through the study of pathological neuroses.

Intuitive Intelligence
Intuition is a very real ability and a difficult concept for some people to comprehend and accept. It is particularly challenging for many scientifically inclined individuals to embrace intuition as a reality. There is this pervasive ignorance that simply because abilities some describe as clairvoyance or psychic are not experienced by everyone, someone who acknowledges clairvoyance as an actuality is somehow irrational or delusional. Talents are not distributed uniformly; the fact that we don't expect exceptional artistic ability in every person we meet in no way prohibits the reality that genius artists do exist. Anyone who's had any interaction with a genuine shaman, or any other individual who interacts with unseen entities, knows that there are indeed uniquely gifted individuals among humanity who display clairvoyance in various ways.

For some, especially shamans, there is a common ability to see unseen entities and converse and interact with them in powerful ways. Many individuals with intuitive abilities can sense others' emotions and perceive detailed aspects of another's character without knowing them. They may be aware of events that will happen in the future or activities happening in the present they are far removed from physically. This ability is documented frequently in writings from various cultures, though little effort has ever been made to characterize such scientifically because these accounts have routinely been flippantly dismissed as erroneous, with rare exceptions.

For example, any medical textbook that describes abdominal aortic aneurysm symptoms will describe that these patients often present to the emergency room with a feeling of impending doom, a sense that something life-threatening is happening inside them, often without any other symptoms. Similar phenomena are documented throughout the world. I've personally observed similar episodes various times in my life, both in myself and in patients and others.

When I was a seventeen-year-old freshman in college, one of my friends developed a severe case of pneumonia. He hadn't eaten in a couple of days by the time I saw him, and I immediately knew he needed to go to the hospital. The way his face looked, a combination of obvious pain and severe malaise, is a look I've seen countless times in emergency room patients. It's a look that's difficult to describe fully with words, but it cannot be faked. It's something you just know, especially when you're used to caring for sick people. And even at that time, years before I'd had any medical training, I knew he needed immediate medical attention.

His roommate helped me carry him to his car, and we drove him straight to the hospital. The doctors were unsure of his diagnosis initially, and a spinal tap was ordered as part of his evaluation. His twin brother, who was living in New York at the time, some 875 miles away from where we were at Grady Hospital in Atlanta, reported feeling severe pain in his lower back around the same time his brother underwent the spinal tap. The pain dropped him to his knees while he was in the shower, and he described having an accompanying feeling from which he knew something was wrong with his brother immediately.

This experience is not unique. Many twins report having a similar experience when their sibling faced grave illness or pain. Some mothers report sensing their child was suffering or had died while away from them. How can these things be explained scientifically? Some would say that these twins and mothers share some psychic bond with their siblings and children, respectively, in layman's terms, but such experiences are challenging to describe scientifically. How can we rationally explain such phenomena? This is not magic. There is a reason why individuals experience such things. What does it teach us about the possible forces, energies, and bonds that exist around us and between individuals? Can these bonds and forces be identified and understood? What might the technological ramifications of understanding such mean for humanity?

When these events occurred, I acknowledged that the feeling my friend's brother described was likely to have happened; I didn't know him, but I see no motive why he would have lied. Back then, I had no interest in understanding what he'd experienced or testing why. It was just another testimony about something that was beyond my scientific ability to explain as a seventeen-year-old. Now I see such an occurrence as yet another frontier for study, an untapped avenue for essential scientific knowledge to advance. These types of phenomena must be examined and understood. I believe they hold the keys to understanding what I consider a novel frontier for technological advancement we've failed to fully grasp. Consider for a moment what might be gained technologically when we understand the how and why of such phenomena. What some humans can do intuitively and through interaction with unseen entities represents a form of technology of the highest magnitude. Such has yet to be thoroughly evaluated for the potential and the danger it merits in an organized and disciplined way. So the line between inherent human ability and the influence of unseen intelligent beings is a thin one, which again lends credence to the merits of and need for significant investigation and study.

In describing and pursuing an understanding of the unknown, words like "magic," "paranormal," "occult," "witchcraft," and "psychic" are often employed. In my opinion, these are archaic terms, and yet concealed within them lie profound truths that have remained hidden from most of humankind. How is science to understand these phenomena? Are we to hide our eyes and minds in denial of the reality of these and other realities simply because we cannot explain them?

Individuals often ignore such realities because they are linked to what some consider evil forces, and for others, these things are simply beyond the realm of possibility. But these occurrences are real. For me, the right question is: What are the forces behind such events? How might understanding how unseen entities work through shamans and others shed light on forces that exist in the universe that have remained hidden from the vast majority of humankind? Removing the shroud of ignorant denial, religious dogma, and mysticism will lead to a far greater understanding of the unseen forces and unseen entities that operate around us. What is the value that such knowledge may reveal?

For believers in Abrahamic religious traditions particularly, the idea of scientifically studying shamans' work with unseen entities might reflexively be rejected and perceived as deeply offensive. But, I ask, what is the moral objection to seeking to scientifically understand works that will be done whether or not they are studied? Individuals practice shamanism and other practices, frequently in secret, and no manner of scientific denial or religious opposition will prevent such practices.

I did a fellowship in addiction medicine when I was a medical student. I was appalled initially after learning a famous and prestigious university in New York City was providing crack cocaine to addicts to better understand the drug's effect on the human brain. Initially, I couldn't understand how such research could be justified ethically. As I grew in my understanding of the study design, I understood exactly why the university ethics board had approved the study. The addicts were selected only after they assured the researchers that they had no intention of attempting to stop using the drug. A pure form of crack was provided that was far less dangerous for consumption than the street version, given that any number of other substances often added to cut street crack were not being used, and the study participants would be monitored continuously. The participants' brains were then studied for the benefit of society at large. The university provided the drug to individuals who would use crack anyway with these individuals' full knowledge and consent.

For those from Abrahamic religious traditions, the thought that unseen entities are a form of alien life might seem revolting. For them, unseen entities are angels or demons, and some believe only demons interact with shamans. The consequences of what such research might reveal will have significant religious implications. If the investigation of unseen entities somehow shows that these life forms do not fit in with the description of unseen entities found in the Bible, Torah, Quran, and other religious texts, religion will experience a significant blow, a crisis of faith. Could it be that what these sacred texts have described as angels and demons are simply a form of alien life that's been completely misunderstood? Such would be a crushing revelation to their religious faith. But if such a study reveals that these holy books accurately predicted the existence of unseen entities centuries before modern science revealed their presence, what might the ramifications of that mean for religion and the authenticity of these holy books, at least to some extent?

Science faces the same crisis from the study of unseen entities. In the modern world, science has nearly entirely ignored even considering the possibility of unseen entities, falling prey to a type of atheistic mysticism fueled by anti-religious sentiment, displaying an ignorance similar to what it often vigorously rejects in religion. How can any human confidently state that they understand everything unseen in the universe or exclude the possibility of life existing in a form separate from organic, carbon-based life forms? There is an enormous price to be paid for science, too, should the study of unseen entities become something that is ever pursued with the intensity it merits. If science proves the existence of unseen entities, what then follows? Questions remain about their origin, how old they are, their intentions, and countless other vital considerations. These realities have tremendous implications. Such a study might very well show that these beings have no connection whatsoever with the heavenly story that proliferates throughout the Bible and other holy books. But what if they do have some relationship with some form of God or creator? Science, too, will be faced with a crisis, as the bedrock of many influential scientific minds is that there is no such thing as God. Herein lies what seems the greatest obstacle to understanding the importance of unseen entities: the reluctance by science to pursue a course that has the potential to disrupt its bedrock atheistic principles.

While it's not a futile course to pursue such cogitation as an academic exercise, the real challenge is not to let biases interrupt that which should take precedence. First, the scientific work is done, and then the ramifications can be fully considered. Whatever the results that come from undertaking the study of unseen entities, pursuing such a course without biases, with the conviction that having a more accurate understanding of the unseen entities surrounding us will ultimately lead to a clearer understanding of the nature of realities previously poorly grasped—we must accept the truth as it is, not as we wish it to be. And whatever comes, may our pursuit always be tempered by a search for a greater understanding of the universe and all that comprises it around us.

I find myself in the unenviable position of being potentially adversarial to both bedrock scientific and religious thought. In many ways, the explicit antagonism of my position convinces me what I am examining must have considerable importance. Finding oneself at the intersection of alienating fundamental paradigms of belief is often the path to great discovery. The study of shamans and unseen entities yields evidence these entities are real beyond question, and as such, science and humanity have taken an enormous step in understanding the nature of the universe and answering vital questions regarding the origin of the universe's unseen realities. Invisible entities have an origin. Whether or not it's related or unrelated to some form of a divine creator, the source of their origin gives us insight into one of the most consequential questions humankind has perpetually pondered. The conclusion that inevitably follows, given that unseen entities are in fact real, is that humanity has made the most extraordinary step in its history in proving God's nonexistence or existence. We are not alone as intelligent beings in the universe.

It will be beyond difficult for some to reconcile their Abrahamic religious faith with the practices of shamans and others who interact with unseen entities. For some, shamanism is a practice of complete moral revulsion. But there are a great many questions that understanding the world around us and all its hidden realities (beings) will answer. Forces and beings that surround humankind constantly are inadequately understood by humanity in general. Some of these beings influence humanity for good and moral behavior; others, for evil, violent, and devious behavior. At times, even the same entity can influence an individual toward both good and harmful behavior.

How individuals are affected by the evil around them, and evil directed to harm them specifically, is something people should understand and have defenses against. One espiritista believes the individuals who interact with unseen entities for harm led to the existence of individuals who interact with similar entities for protection and healing. The reality is that some individuals use their ability to communicate with unseen entities solely for the service of humanity benevolently. How are we to think of them? Should they be considered outcasts, many of whose acceptance of their ability was coerced over years of harassment? Is an individual who lacked the freedom to accept or reject something they were born with someone who should be condemned? That is an ethical question worthy of the most profound consideration. Society must confront such if we are to work in harmony with shamans and others who interact with unseen entities in an effort to understand these entities fully. These individuals are the experts in such practices, and it is prudent to consider their input when planning and developing research studies. To deny the existence of unseen entities and shamans' abilities to interact with them, and the intelligence and power that unseen entities manifest, is to deny a reality that exists around us.

In my concept of God and the universe, I believe understanding the true nature of the universe and all it contains is a path to inestimable scientific knowledge. To love is to live in harmony with all that surrounds us, to live with an awareness of all that both exists and is real. To embrace that reality, its good and evil, its ugliness and its beauty, as that which it is and not what we hope it to be is to stop hiding our eyes from the truth that is. The reality of all that surrounds us, things hidden and beautiful, evil and ugly, and understanding its ramifications for all beings and the environment around us moves us a step closer to what some might consider the divine within us. When we live choosing the better aspects of who we are and the gifts we have, the purer side of man's nature and consciousness, we use all we are in humanity's service. It is not loving or reasonable to hide our eyes from that which is true, though it might be considered evil or ugly. Understanding the unknown and the poorly understood, and using that knowledge to reaffirm our commitment to what is right, is a beauty that dignifies our hearts. What we are, what we're capable of, and humanity's place in the universe are hidden in the knowledge, understanding, and deeper awareness of all that remains unknown around us, yet to be discovered and fully understood.

Cultural Superiority

When considering the reality of individuals with intuitive intelligence, I asked myself a vital question: why is it that the Western world so vehemently denies the existence of unseen entities when other nations, particularly African countries and other nations with larger populations of Indigenous peoples, regularly acknowledge their existence? If, for example, you ask most Native Americans, Cubans, Africans, Aboriginal Australians, or persons of Inuit descent if unseen entities exist, they'll reply immediately, as if acknowledging a matter of obvious truth, that such entities do exist. This duality engenders a significant inquiry: why such disparity in thought?

Often Americans, particularly certain ethnic groups within America, think of racism as a thing of the past. There's often this notion that racism ended with the end of slavery or the first Black president's election. Racism, and the powerful effect of prejudice, can be observed throughout history, not just in America, but in every part of the globe. It has produced tangible and long-lasting effects and, in many ways, has shaped the course of history and the paths of nations for hundreds of years. The impacts of colonialism, the Crusades, religious indoctrination, slavery, apartheid, India's caste system, and other forms of prejudice and bigotry still linger today. Its consequences have crippled humanity's ability to achieve its optimal potential. The denial of the existence of unseen entities reflects the subtle but powerful effect of prejudice. Nearly every known Indigenous culture that has ever existed, including those still present today, describes these entities' existence. Yet, Western civilization has clung to the belief that these numerous peoples are pitifully ignorant, enslaved to some mass delusion. The very refusal of society to accept the innumerable accounts of these predominantly nonwhite peoples displays how prejudice has the power to supplant the optimal destiny of humanity. Had such accounts of these entities predominantly come from nations and peoples of European descent, they would've long been met with the assumption of accuracy, and the subsequent study of these entities would have changed the course of human history long ago.

Cultural superiority is an ongoing problem within the educational framework of America. I cannot speak of other nations, but it seems something similar persists throughout the world, as racism and prejudice appear as alive today as they've always been, and the wave of extremist right-wing political movements around the globe displays its ongoing not-so-indolent presence. People's thinking is often reflected in the literature they espouse, and educators are no different. When I was in high school, I was mandated to read books from writers whose perspectives and fictional narratives were nearly entirely White and Eurocentric. The entire curriculum of four years of English, in a school more than 50 percent Black, involved not a single novel, poem, or nonfiction book written by a person who was nonwhite. Instead, I was inundated with books like *The Great Gatsby*, *Huckleberry Finn* and its prolific use of the word "nigger," *To Kill a Mockingbird*, Edgar Allen Poe, The Puritan, Planter, Pilgrim Period, and the like. Whether this was an intentional oversight or just the ignorance of racially aloof educators, I cannot say. In thinking back, my previous statement was incorrect. I had to read one book by an African-American writer, Richard Wright's *Native Son*, a story about a poor Black boy who accidentally murders a White woman.

Works by Shakespeare and Sophocles, pieces like the Iliad and Odyssey, and other established masterpieces are appropriate works to be enjoyed and studied. Still, they are by their very nature antiquated and racially, extremely homogenous. The world is full of brilliant writers, artists, musicians, and other geniuses, whose works and ethnicities are as varied as the diversity of their genius. By what more significant gesture can we demonstrate what and who we value as a society than by the manner in which we educate our children? By the very nature of who and what we exclude from our education, we also demonstrate who and what our society values. For years, a segment of educators and others have complained about the cultural biases within standardized tests like the SAT and ACT. These tests have their failings. The real problem is in a system of education that exudes cultural biases at every level, and that these tests fail to evaluate students in the areas that matter most.

How do you measure a person's integrity, honesty, empathy, leadership ability, determination, ability to overcome obstacles, creativity, work ethic, spatial intelligence, commitment to environmental conservation, interpersonal skills, and other qualities essential in the modern world and beneficial for our society? When we consider these vital qualities and abilities and look at how we evaluate students, it is clear such antiquated instruments focusing principally on factual knowledge and reading comprehension are ineffectual. We have completely ignored other factors that are far more critical indices of character and more effective in predicting a person's likelihood of success. These instruments are the tools of an antiquated educational system that has failed to instill in humanity the qualities needed to maintain a harmonious balance within our society and with the natural world. For as Mays's quote thoughtfully illustrates, "we have no guarantee that when we train a man's mind, we will train his heart; no guarantee that when we increase a man's knowledge, we will increase his goodness. There is no necessary correlation between knowledge and goodness."

Education Void of Moral Instruction

When we take a closer look at our educational system, the entirety of its focus, with the exception of, say, religiously inclined private schools, is on instilling factual knowledge and reading comprehension. Simply put, we focus on the mind and not the heart. Not surprisingly, the effect of this approach is reflected in the lives of the macrocosm of society, a society that focuses on success defined and measured by wealth and educational attainment, essentially acquiring things, void of any moral, ethical, or environmental conservational principles. Is there any wonder we see our planet in ecological decay and a world fractured by war, moral decay, and the persistence of gross degrees of inequality? If our children are not educated in the values that are principally needed for sustainable existence with our planet and our species, how could they manifest these values as adults?

In decades past, efforts for moral instruction were made within the US educational system, but always interwoven with religious indoctrination—prayer in school and the reading of the Bible, for example. Parents appropriately withdrew from such efforts, as religion is a diverse and deeply personal choice, and the state should not impose religious rituals and standards on its populace. Religiously minded people, particularly Christians, frequently believe only Christians are people with morals, but this is far from the truth. Practically every religion advocates ethical principles. Even those without belief in God can often manifest ethics and moral behavior of the highest degree. At times, those who present themselves as religiously devout individuals manifest behavior that is far from moral and ethical. It is possible, and it must be encouraged with the most extraordinary vigor, to instill benevolent values within our children as a part of their educational experience. Encouraging benevolence, including principles like honesty, integrity, proper work ethic, empathy, generosity, tolerance, impartiality, and other universally recognized beneficial qualities, in our children need in no way be connected with religious indoctrination and instruction.

A child's mind is a sponge, and children understand even complex concepts of fairness, equality, impartiality, and benevolence at a very young age. Watch any child playing, and ideas of discipline, sharing of toys, and fairness are concepts they easily understand. As prejudice is a learned pattern, so qualities of benevolence must be instilled. It is a failure of considerable consequence to delay such instruction in favor of exclusively limiting education to the mind.

It's easy to leisurely criticize aspects of society without offering any solutions. And while it's challenging to address all the complex problems in society, I have salient thoughts on the path forward to bring more balance and harmony to humanity and our fragile environment. A lack of empathy and awareness has contributed to the development and sustenance of a significant portion of these issues. Still, education is a powerful tool, and it is the only real pathway to remedy the enormous challenges our species and civilization now face. As technologies push society forward, and globalization, artificial intelligence, and other aspects of external technological developments drive us further away from principles of benevolence and sustainability, the true path forward rests in embracing the philosophies and characteristics of cultures long ignorantly thought primitive. Let's consider some ways the modern world can learn from the wisdom of Indigenous peoples.

The Mind and Heart of a Healer

Out of all the interviews and experiences gathered in the process of compiling this work, the one that follows here is perhaps the experience that evolved my perception of traditional healing, science, and ethics most profoundly. It thoroughly reflects the beauty often contained within various forms of Indigenous philosophy and other religious thought.

My home in Havana is simple and modest and covers all my needs, and I live a simple but happy and contented life. I've met and helped many musicians and artists trying to climb in their professional lives. Many of them have either become very famous or were already famous when they came to me, and I have helped them to see their paths more clearly. I have not traveled to any foreign countries. I've never left Cuba and do not need to or want to.

How would you describe the spiritual realm where spirits reside?

There is no way to describe the spiritual kingdom. It can only be described in terms of sensations, with the power of energy, and with the peace and tranquility that emanate from it. But you can also feel the weight on the shoulders of humanity because of their problems. Energy can be felt in different ways: in physical strength, in psychological strength. Bad energy causes general malaise.

What do you see as the future of humankind?

The future of humanity is as unpredictable as the actions of any individual person. Man has a killing nature and is destroying the natural world, and nature is where the saints feed and where there are the means to heal all of humankind's sicknesses. Humans should live in humility and be fearful of God's strength, as he is the only one capable of creating and eliminating everything bad around us.

What is the significance of having an altar?

The altar is where we keep the images that represent each of our saints. It is where we communicate with and feed our saints, where we ask for and receive all the healing and miracles.

What would you like people to understand about the saints? What do you think about technology?

We do not choose the saints who come to help us. They choose us. I would like humanity to know that the saints can help us with every problem in our lives. Tobacco, sweets, and other offerings have been made from the time of our ancestors in payment for the good works that the saints perform for our benefit. The opinion I have of human technology depends on how they use it for their benefit or against it. What most annoy a santero are lies and false words. What most annoy the saints are the false prophets who perform actions in their names only in order to obtain their own benefits.

Is there a usual way a person grows in ability and power as a santero?

There is no usual sequence of how a person advances in their knowledge and skills as a santero. From the time the person is young, and a saint chooses him to one day become a santero, that person can feel the presence of the saint who has chosen him. In time, this communication of the saint's presence becomes so frequent that the person agrees that this saint should flow and communicate through him. When the person enters a trance or is in transition, it is only a means for the saint to communicate with the outside world. While in transition, the santero will often be granted special abilities to perform different jobs. Over time, the person gets used to it and allows the saint to flow with more force and intensity.

Are some santeros more powerful than others?

There is no saying that one santero is more talented than another. All saints work for the same purpose, the difference being how a person lets the spirit flow and work through them. Some do it by faith or by strength of will to help others. These are the real santeros. Others do it with falsehoods or use the gifts they have to harm. They only let the part that suits them for their own benefit flow. Others are false spiritualists: people who do not have the gift but can imitate. They are the ones only seeking an economic benefit.

There are also santeros who live in opulence and luxury, a consumerist life. But as the Bible says, nobody leaves this life without paying for sins committed. The true santero, the one who truly possesses the gift, lives in humility and near the mountains, since this is the force that feeds him. Everything in Santería has an African origin. The true santero does not need the consultee to tell him their problems. Just by observing the person, a santero can see their problems and needs, and the diseases they suffer. There is no guessing to get the diagnosis right or doubt about how to solve such problems. The false santero plays word games and uses conversation to obtain information that is not always effective when giving the diagnosis to the consultee.

How can santeros sense when there is someone else near who works with spirits or saints?

When a person who has some hidden influence or power approaches him, a santero of high experience can feel the energy emitted by that person. It is something inevitable and inexplicable, like two magnets drawn together or a chemical attraction comparable to when a man and a woman who have never seen each other feel an instant attraction.

What do you think is the best way a person can live?

The best way a person can live his life is to be at peace with himself. The most important thing is that you become and also recognize in yourself that you are a good person. Most successful people take advantage of people. They may go ahead and prosper, but they will not know true peace and happiness. How long will that last? Twenty, fifty, sixty years? The life of a human being on Earth is only one-thousandth of a second of what it takes to come and go in eternity before God and the saints. We will be accountable for the short time that we lived, and these will be the ones in charge of condemning us to live in eternity for the actions we have committed. The road to hell is wide and spacious, and the road to paradise is narrow. We should help others and worry about the problems and needs of our family and friends and then think about ourselves last.

How do you feel about the Bible?

The Bible is a book guided by God but made by men. It is well guided but poorly elaborated due to the imperfections of men. It is a book that requires interpretation, which is why many religions have used it to achieve the ends they desired.

A Missionary Witnesses the Paranormal

My wife and I first met when she was leaving the gym and I was going inside it. We went on one date, and then afterward only corresponded through letters since I left shortly after to Africa for my missionary assignment. Three years later, we got married when I returned to the States on a break from my assignment.

I have known a deep fondness for science and nature since I was a child, so I moved to southern New Mexico to study biology at New Mexico State when I finished high school. The order I found in nature convinced me that there had to be a God, as I reasoned it was unlikely that the special order in the natural world surrounding me occurred without some organizing intelligence. That thinking led me to begin searching for God, and I started reading all sorts of religious texts from around the world. I remember praying, imploring God to reveal himself to me if he existed. Though I grew up going to a Catholic church because my dad was a Catholic, he grew angry with the church when I was young and stopped going to mass. Despite my mom being Lutheran, she continued taking me to mass after my father stopped going, but I never was a firm believer as a child.

Though raised with that foundation of being around the church, I was far from a real Christian in my early college days. I believed in God, but I wasn't sure if there was really a religion that God favored more than any other. I enjoyed living a sinful life and was no stranger to substance use, so though I had a measure of faith, my life at that time was certainly not one worthy of imitation.

Shortly after beginning my college studies, while I was living in the dorm at NMSU, two young men knocked on my door and asked me if they could study the Bible with me. I agreed, and one of my close friends sat with me as they began sharing various scriptures from the Bible with us.

They showed me Romans 3:23, where it says that all have sinned and fallen short of God's glory, and John 5:24, where it says that whoever believes in Christ has eternal life and will not be judged, but has crossed over from death to life. They asked me if I believed those scriptures were true. I did. I became so convinced that what they were sharing with me was the truth, I left and gathered other students in the dorm to sit in the Bible study with me. By the time I gathered everyone I could find who was willing to sit with us, my dorm room had eighteen of my classmates listening while these two young men sat teaching us from the scriptures. That night, two of us accepted Christ and got saved.

When I finished college, I began working for New Mexico Fish and Game and would spend hours alone in the mountains and forests observing God's creation more fully. In my downtime, I would read from a small Bible I carried with me and memorized countless scriptures that strengthened my faith and deepened my love for Christ. I felt a burning desire in my heart to share the beauty of the Bible's message with others. After a year and a half working in the beautiful landscape of New Mexico, I headed to Portland, Oregon, to begin my studies at seminary.

The year at seminary flew by. I progressed well in my studies and was an earnest student. Each year, the school features a missionary week, and missionary organizations from across the country visit and encourage students to consider serving as missionaries throughout the world. I became convinced that missionary work was the right choice for me and invited one of the visiting presenters to have coffee to understand better some of the options available to assist me reach my goal. As we sat and talked, he mentioned a program in Nigeria that employed expats to teach in a boarding school. This opportunity ended up being perfect for me. It would allow me to begin missionary work right away and not depend on a church to carry the financial burden for my assignment.

Though my mom was less than enthusiastic about my choice to go to Africa, she figured there were worse things that I could be doing with my life. Shortly after finishing seminary, I left for Nigeria, excited and convinced that the Lord had heard my prayer and was blessing my desire to spread the gospel. The boarding school offered housing for teachers, and I stayed on campus as I began my new life there. There was no air conditioning, and the conveniences I had grown used to in America were not a part of my routine anymore.

The days started early at school. I would teach from 7 to 10 a.m., and then from 10 to 11 a.m., a hot breakfast was provided. Classes resumed from 11 to 2. At the time, Nigeria was experiencing a financial crisis, and frequently, teachers would strike as the government failed to pay us. There were many times when classes would be interrupted for a few weeks, and the administration would pay me with a large bag of rice. The conditions were less than ideal, but the Lord was blessing my ministry, and that made continuing in the assignment easier despite the hardships.

Over the first few months, I had twelve Bible students. We would meet in the afternoons and on weekends. From the very beginning, my approach was to train disciple-makers, and I implored my Bible students to take their studies seriously. By the time I left Nigeria, my Bible study group had grown to sixty students. Only a year and a half later, those first twelve students began conducting their own Bible studies.

I left Nigeria for the Ivory Coast after feeling like my efforts would be more impactful there. Though I enjoyed Nigeria, and my ministry was proving fruitful, most of the people there owned Bibles and had at least heard of Christ. It was my goal to preach to people who had never really known anything of Christ. I reached out to my missionary office, and they helped direct my efforts toward the Ivory Coast, where the circumstances were more in harmony with the work I was hoping to do.

When I arrived in Africa, one of my new local associates warned me not to cut my hair outside or leave any personal belongings where they could be stolen easily. He mentioned that children sometimes steal such things and sell them to shamans and witches, who use the items to do more powerful witchcraft. I initially dismissed his warning, but it wouldn't take long to appreciate his words fully.

As my time in the Ivory Coast grew, so did the opposition to my missionary work. The area was predominantly Muslim. Though the town and surrounding villages were open to our work, a militant minority among the Muslim clergy deeply resented us. I was warned that the clergy had been paying the shamans and witches to curse us, but the rumor spreading around the town was everything that they were doing to try to curse us would return to them.

When a person is a true Christian, the Holy Spirit abides with them, and no demon can possess them, and no curse can harm them. I knew that the Lord was with me, protecting me and blessing my work, so I was never afraid of any curses. But I did have a few experiences that convinced me that the shamans' and witches' work was very real.

Many times, villagers and others would come to our home or church asking for our help with spirit-possessed children or other family members. In most instances, our assistance was declined after we explained that the person affected would have to become Christian to prevent the demon from possessing them further, as it says in the scriptures. But there was one instance worth describing.

One day, the parents of a child from a village some distance away came to us asking for assistance for their teenage daughter, who had become possessed by a demon while washing clothes at the river adjacent to their village. They asked that we come and pray for the girl, so one of my fellow missionaries and I drove several hours to the village. When we arrived, there was a Muslim shaman in the village, and we recognized the shaman immediately—we'd seen him in other villages. We asked the parents what they wanted to do because we would not both be able to help the child; they would have to choose between the shaman or us to help the child. The village elders "hung head" and discussed among themselves before deciding that they preferred to let the shaman work to help the child. I asked if we could observe the shaman's work, since we had driven so long to see the child. The village elders and the shaman agreed.

We entered the hut where the child was sitting. A fire was burning low in the hut's center, and a child sat motionless at the rear of the hut. The girl was sedated from some herbs that the villagers had given her in the time before we arrived. The shaman brought a small, lightly colored wooden board into the hut with him. He took some of the charcoal from around the fire and began writing verses from the Quran on the board with it. He then took some water, poured it on the board, and collected the ashes and water in a pot as the water rinsed off the words written in ash on the board. He wrote a few more scriptures and repeated the process, again collecting the ashes and water in the pot as he poured the water over the board. He then told the possessed girl to drink the water and ash mixture. She did, and as soon as she finished, the girl began screaming and ran out of the hut.

The girl's brothers and a few other young men were sitting in the bed of a pickup truck outside the hut and began driving to follow her as she began running toward the river where she was first possessed. The men described going thirty-five miles per hour in the truck following the girl as she ran but still not being able to keep up with her. After returning to the river, the men reported that the girl screamed and instantly returned to her normal state of mind. When we next saw the girl after they brought her back to the village, she was wrapped in a blanket, shivering, but talking and acting in her right mind.

There was another experience that bears sharing, but for many, it will seem beyond belief. A large adobe wall surrounded the compound where I lived during my time in Ivory Coast. One evening, late in the night, I heard our dog barking. It wasn't uncommon for our dog to bark at passersby periodically. Still, when he continued for several minutes, I grabbed the bat I kept under my bed for protection and went into the front yard to see what all the commotion was about. On the adobe wall near the front of the house, a huge barred owl sat calmly while our dog continued barking and jumping up, trying to bite it. I approached the owl from the side, hitting the side of the wall with the bat, hoping to scare it away. The owl just looked at me calmly and didn't move. It was not until I got close enough to be able to hit it with the bat that it flew away. I went back inside the house, puzzled at the owl's behavior. This behavior was not appropriate for an owl. Usually, an owl would be afraid of a dog, let alone an approaching human.

The next day, I had a meeting with our translator, who was helping me translate a movie we had been showing villagers about the life of Jesus. As usual, we began our work, but I couldn't resist telling him about my experience with the owl the night before. After I told him, he started sneering with laughter. I said, "What? Why are you laughing?"

"That was no owl, my friend. That was a witch."

"What?!" I said in disbelief.

"I bet you I can tell you exactly what time he appeared. He was there at four a.m., right?" He was right. "That is the time they are most active; they can fly all over the world. Everyone around here knows what they can do. He was trying to curse you, but because you're a Christian, he couldn't enter your yard."

In Africa, everyone believes in spirits. There is often this extreme view that nearly every illness involves some sort of curse from someone. Many times in Africa, I took children and patients of every other age to the hospital after convincing the person or their parents that their illness was a physiological disease and not the result of witchcraft. Admittedly, there were times when it was hard to tell the difference. In the Western world, it's the other way around. Science is king, and very few acknowledge the existence of spirits at all. Christian missionary ethnographers describe what they call "the excluded middle," a concept that they believe the world should move toward. It's this idea that there is a middle ground that is often ignored, where society rationally embraces the existence of unseen entities and yet does so while maintaining that there are physiological causes of disease and problems.

I am a biologist and a Christian. It is a duality I resolved many years ago. My scientific understanding and experience did not weaken my faith; it has been a tremendous asset in my work as a missionary and disciple-maker. I am a rational thinker, yet I acknowledge that there are spirit beings that exist exactly as described in the Bible. But as a Christian, I feel no fear of evil spirits. The blood of Jesus protects all Christians, and I feel comfort in knowing my family and I are safe with the protection of Christ.

A Santera Remembers

I was born on a farm in Baguano, Cuba. My family was very poor. My dad worked as a lumberman, cutting down trees in the woods not far from our farm. From the time I was very young, perhaps four or five, I would see family members and friends of my father coming to my house. They always went to the backyard, where my dad had a small shack. I could hear them speaking in a strange language, but I never got close enough to see what they were actually doing.

Time passed, and when I was around ten, I realized my dad was either a santero or a palero. When I grew older, I learned he was actually a skilled santero. When he could no longer make a living chopping wood because of his health, he dedicated himself to helping people as a santero. People would come to our house. Although he wouldn't charge them anything, they always brought him animals or sometimes food or money. We always had enough to live, but we never grew out of poverty. According to the comments of many people I met as I matured, my dad was very good at what he did as a santero.

I grew up in that world, surrounded by santeros, devotees, and spirits, but I didn't practice anything. I was just a kid in school during the time my father was most active as a santero. Whenever there was someone with a disease or people came to him with other problems, my dad helped them with his work and the help of the saints.

Things changed when I was in my early teens. My dad became very sick, and one day he took me up to the top of the mountain near our home. We brought with us some snails, tobacco, tree branches, and rum. He told me he was going to die soon, but that he would leave me his gift before he died. He started doing rituals with me using the things we had brought with us up the mountain. When the rituals were finished, he told me he had transferred his gifts to me. But I didn't understand at the time what he actually had done.

He died soon after that day. I didn't see spirits or sense things right away, but when I turned fourteen, everything changed. I started to feel the presence of things around me and see spirits. I tried not to pay attention to what I was seeing and feeling, but as time went by, my mental health worsened. I could not make any plans or find success in my life. If I had a boyfriend, he would leave me for no apparent reason at all a few weeks after we started dating. I eventually dropped out of high school because I couldn't concentrate. The spirits were harassing me too much. Little by little, I fell deeper and deeper into depression. I almost became crazy. I spent more than six months in a hospital with actual insane people, but I knew that I wasn't crazy.

One day my aunt, one of my father's sisters, told my mother that she wanted to take me to visit a very good curandera she knew. This curandera lived in Guantánamo. My aunt, my mom, and I rode to see the curandera in a horse and buggy. When I was with her, a dead man appeared to me and told me, "If you don't work it, the gift you have, it will kill you little by little." Shortly after that, my mother took me to the house of a santero. It was he who taught me how to channel everything I had.

As I learned everything related to Santería, my life gradually started changing. My depression went away, and I started to work as a santera. I still remember my first consultation. My cousin came to ask me for help. She was suffering a lot from her husband's infidelity. He had mistresses, many of them, and cheated on her all the time. I asked her to bring me a pair of his boxer shorts and a photo of him.

I consulted with my saints and did a ritual involving his boxers and the photo. In time, he began to have an eruption in his skin, which developed into an infection, and pieces of his skin began to fall to the ground. In time, it spread more and became a bad plague. He could no longer be with other women. I didn't feel bad about what I had done. He had brought it on himself by what he was doing.

I've helped many people with all sorts of problems over the years, but only through the power of the saints. Once a man came to my house whose boss wanted to remove him from his job but couldn't fire him right away without a reason. I went to my altar and went into a trance. I don't remember what the saints told him or what happened during the consultation, but later that day I went to catch the bus, and the man who had consulted with me told me I had told him not to worry, that his boss wouldn't be among the living much longer. Three months later, a truck loaded with cane killed him. The man came back to see me when the accident happened, worried about what had happened and worried he had caused his boss to die after seeing me. One of my saints was nearby and told me to reassure him that we had nothing to do with his boss's death. I had simply told him what was already destined to happen. He left my home and was grateful to the Lord.

I have a cousin who is a doctor and was chosen to fulfill a medical mission in Venezuela. She was excited about the trip, mostly because she had never left Cuba before. But a couple of weeks before she was scheduled to leave, she was notified that her trip would be delayed because her spot in the mission had been given to another doctor with more seniority. She was more than disappointed. She came to my home, hoping there was something I could do to help her.

I consulted with her at my altar, and the saint told her not to worry. In a short time, she would leave on a medical mission. She came back to see me twenty days later and told me that the other doctor who had been sent to Venezuela before her had been the victim of a robbery and had been beaten and threatened. The other doctor had decided to return to Cuba; my cousin had been immediately asked to prepare to take his place, and she would be leaving shortly. She accomplished her mission without incident and bought a house in Holguin when she returned. She is doing very well.

I have a nephew who is a farmer. One day he came to my home distraught after one of his bulls was stolen. That bull was everything to him—it was his whole livelihood. It was yoked to another bull every day, and they were all he had to plow his field. The night after he was robbed, he went to the police, but they couldn't find the bull or the person who'd stolen it.

When my nephew came to see me, one of the spirits said to me, "Hurry up! They have it in an old, unused street, and if you don't hurry, they are going to sacrifice it." He didn't believe what the spirit told him through me, but my sister was also there, and she told him to go at once.

He went with my brother-in-law and two of his friends to the place the spirit had told him. When he arrived, he saw two young guys running out of one of the abandoned buildings near the end of the street. When my nephew entered the place the men had run from, he found his bull tied up and ready to be killed. He rescued his bull, then came back to my house and thanked me for how the spirit had helped him.

One of the most difficult cases I ever had involved a young married couple. This couple was having problems in their home and in many aspects of their lives. They were having problems getting along and having other problems with work, too. When I went to my altar and consulted with my saint, I learned that their house had a hell of a disease. A dead spirit came to me and told me someone had cast a powerful Voodoo spell on the home. That wicked person had symbolically buried the couple in the yard of the house by taking dirt from the cemetery and doing a ritual with the dirt and other objects to cause harm to the couple. I could see through the altar that the container holding the bad work had been buried in the garden of their house, but I could only take it out at noon on the twelfth day of that month.

When I told the couple how the bad work had to be removed, they became distant and detached. I could sense that they were afraid and felt ashamed because, at that time of day, everyone would be seeing me coming to their home. I had become well known in the community by this time, and they did not want people to see me at their home and know that they were having a problem. They left my home and made no arrangements for me to visit their home to remove the container.

A month passed, and the couple came back to my house more distressed than the first time they visited. Their youngest daughter had become very sick, and the doctors did not know what to give her. The couple insisted I come with them. They took me in an expensive taxi to their home. I didn't know if I would be physically strong enough to leave the home after removing the work because of how strong the bad work was. When I arrived at the couple's home, I went into transition, and a spirit began operating within me. Later, my husband and niece, who had come with me, told me that after I went into transition, the spirit snapped my clothes into pieces and I began crawling nearly naked around the backyard. I made my way to one of the bushes, where there was a large mound of red ants. I began digging through the mound of ants, and as I did, the ants began to cover my body. My husband told me the more I dug; the more ants covered me. He became afraid and wanted to stop me from digging, but then the spirit turned my head and spoke to him in a scary voice: "I am her guardian; nothing will happen to her." My husband left me alone and let me dig. After half an hour of digging with my hands, I pulled out the can that held the evil work.

When I did this, the ants began falling off my body. My husband and niece took me out of the yard. I was almost naked, still holding the can, and nearly dead. I asked them to take me to the nearest river. They rushed me to the taxi, which was still waiting, and took me to the river. When I stood in the river and opened the can, I fainted, and everything went dark around me, but the river's current immediately took the can and everything in it away. I would have drowned, but my husband reached in and pulled me out of the water. It was a difficult experience, but a necessary one. I woke up nearly naked and in tremendous pain, and that is what I remember most. To this day, I am still very good friends with that couple, and their little daughter became my goddaughter.

Another time, a woman came to see me because her son could not stand up. He had to do everything lying down. My husband and I went to see him. He was a very good-looking young man. A spirit told me the boy needed to be given a specific remedy. I made the drink, and when his parents gave it to him, he began to vomit. The remedy partly worked. He sat up but still could not stand, so I started to pray over him. Then he suddenly stood up and went into the bathroom, where he began to have a long urination. The smell of it was pure rot. I knew from the smell that the bad energy was leaving him. Years later, he came to see me because he had gotten gonorrhea. "That's a job for a doctor," I told him. "I can't do anything about that." He just laughed.

My husband was only half in love with me. One of my spirits told me he was having an affair, and I told him I wouldn't let him play with his daughter until he stopped cheating on me. A few weeks later, I saw him, and he seemed very sad. I asked him what was wrong. He was embarrassed and told me he couldn't get his penis up. I laughed and told him that was a punishment for his infidelity, and the next time he was thinking about cheating on me, he should remember that.

An experience that always deeply impacts me is a ritual I do known as a *Change of Life*. In this ritual, the life force of a person near death is given to a spirit, usually in exchange for healing someone young who is very ill. The ritual is done using a spirit and dirt taken from a cemetery. Two dolls are made: one that contains the dirt from the cemetery and one with nothing inside it.

One *Change of Life* experience I'd like to share happened two years ago. It was an experience I'll never forget. My sister-in-law's nephew was very sick, and the doctors had given him a very short time to live. I went to the cemetery to find a spirit to help me with the ritual, then made the two dolls and went with my family to see the boy. When I saw him, he was indeed near death. Time was short if there was any chance I would be able to save him. Next, I went to find someone who was terminally ill. There was a very ill man who was eighty-four years old and ready to die. He was willing to be used in my ritual. I looked for a few herbs and other things I needed for the ritual, and the next day, with help from a woman who cleaned in the hospital, I did the work at one o'clock in the morning after sneaking into the hospital. After passing the doll filled with cemetery dirt over the elderly man, I went to the boy and passed the clean doll over his body. I felt all the bad energy and sickness flow through my body as I passed the doll over the boy. I finished around four in the morning. When I finished at the hospital, I went straight to the cemetery to deliver the life of the old man to the spirits. The boy immediately improved and was discharged from the hospital shortly after. He now comes to see me on the day of St. Lazarus, still very grateful to me and the saints.

I can't believe all the work the saints have done through me. My saints have saved marriages (and at times also ended marriages of many years). They've healed people sick to the point of death, helped people in wheelchairs to walk, and healed blind people. I have also helped many people possessed by bad spirits. I've done works that hurt and even killed evil people and done *Change of Life* work that uses the life of a sick person voluntarily given to heal others. I know that it is not me as a person that does this work; it is work done by the saints and the spirits my father left me. I have no remorse for what I have done. It is a gift that was given to me for both good and evil, and all I have done through my body, I do only through the help and power of the spirits, so I feel no guilt. For decades, people have come to my house with their problems. I have many beautiful and bad secrets and memories of the work I have done. But I know without a doubt that this is the work I was meant to do in this life. When I go to a better life, they will judge me for what I have done in this life, but I have no fear.

A Santera Protects Her Daughter

*AC**

The account below details the occurrence of an apparent stroke (cerebrovascular event) after a prayer was offered by one healer to defend her daughter. It will stretch the reader's credulity. You can make up your own mind about its accuracy. It was observed by twenty-plus eyewitnesses."You don't mess with witches, bro." That was a straightforward proclamation of one of my Mexican friends on the far East Side of El Paso years ago. No stranger to the spirit world himself, he didn't have to tell me—I already knew it well. The account below was just one of many that opened my mind to the power of unseen entities and the humans who've cultivated a relationship with them. It is a well-known event in this coastal Cuban community.

I come from a long line of santeras. My grandmother, great-grandmother, and great-great-grandmother were all santeras. And, like them, I was born with the gift. But I didn't want it. My great-great-grandmother was an African woman who married a Spanish (White) man in Cuba. Their daughter, my great-grandmother, was brown-skinned, lighter than her mother, but still had kinky black hair and African features. She lived to be 115 years old.

My great-grandmother also married a light-skinned Spanish man. His name was Roberto, but everyone called him Chicho, after a cartoon character in a Cuban comic book. The character was miserly, and so was my great-grandfather. He knew every way there was to save a penny. He was a farmer, and he also loved making cigars in a little room on the side of our house. When I was young, maybe four or five, he would make cigars with me in that room. It's one of my earliest and fondest childhood memories. Chicho showed me how to cut the tobacco leaves and dry them. Then he would place the dry leaves in a little metal machine he had made and roll them with a little wheel that had a handle. Afterward, he would hold the newly formed cigars next to a small candle and burn them to finish the rolling process. I remember talking to him for hours at a time. It wasn't until I was older, my mother told me he had died long before I was born, and the man I thought had taught me to roll cigars was actually a spirit.

My great-grandmother spent a lot of time in the backyard of our house. She would use my grandmother, as later my grandmother would use me, during *bembé* and other rituals to channel the spirits. My home was a popular place. People would constantly come to see my great-grandmother and grandmother. Though she was about 110 years old when I was born, my great-grandmother could still dance and move like a woman in her sixties. When my grandmother wanted to sew, she would bring her thread and a needle to my great-grandmother to thread the needle for her. My grandmother's hands weren't steady enough to still do it, but my great-grandmother's still were. My great-grandmother looked much younger than her age, and it was only after she died I really understood how old she was.

My great-grandmother used my grandmother during rituals to channel spirits. There were usually dozens of people at the house. My grandmother, in her seventies at the time, walked, talked, and moved with her father's mannerisms. I didn't really understand what was happening, but I didn't feel afraid. Everyone was there at the house, no one was afraid, and nothing felt unusual or out of place about the ceremony. I just sat in awe, watching what was going on.

When I was a child, I liked to be adventurous. I would run everywhere, like a Cuban version of Forrest Gump. My family and others in the community began calling me Pilín, and when I was running in town, the older men and women always shouted, "Pilín! Pilín!" since that's what my dad called me. I had these ugly government-issued shoes, and when I had worn them for just a little while, the fibers would burst out their sides, hurting my feet. I would complain to my grandmother, and she would say, "Don't worry, child. When you're older, you will have so many shoes you won't know what to do with them all."

Born in the time before Fidel Castro took control in Cuba, my grandfather was an accountant and my grandmother a hairdresser. Grandfather purchased the land that held the local airport and renovated the building into a six-bedroom house. Fidel came to power in the middle of the renovation, and my grandfather couldn't get the materials he needed to finish the house because of how the revolution affected the buying and selling of goods. Instead, he and his brothers foraged in the local area and used tree branches, driftwood, and twigs to patch the expansion together. It had been started with wood and brick and finished with sticks and twigs. When it stormed—and it storms a lot in Cuba—I'd watch Grandfather chase after the twigs that made up the walls and the roof. Every twig had its place, and my grandfather was a master of keeping all the twigs in their right places.

The floor of the house was entirely made of dirt. When it rained, it got muddy, and when it was dry, it was dusty. My grandmother would carry an old can and sprinkle water on the dry ground in the morning so the dust wouldn't blow everywhere during the day. She used one room in the house just for her Santería. There was a larger wooden table covered with candles and statues in that room. There was also a wooden frame extending from the table up to the ceiling, and wreaths of red and white flowers hung above and around the frame. Three other rooms in the house went unused after my mother's sisters and brothers moved out. My aunt was the youngest, only sixteen, but by the time I was old enough to remember, she was already married and was the last of my grandparents' children to leave the house.

The kitchen was my favorite room, the room I remember the best from my childhood. My grandmother would wake up early every morning and boil lavender and other herbs on the wood-burning iron stove. It made the kitchen smell lovely, yes, but the whole house was filled with its fragrance. We'd use the fragrant water left over for bathing, herself and me, because the soap the government provided would break me out in a rash.

Beside the iron-burning stove sat a small three-legged circular table with three chairs around it. This was where we would sit and have breakfast and dinner together. There was a single window on the far side of the kitchen, next to the stove, and my grandmother sewed linen curtains by hand from material she'd gotten before Fidel came to power.

My grandmother was still beautiful at seventy-five—*vieja pero bueno*, or "old but good," she would often say. She was seventy-five when I first remember her, but she looked at least ten years younger. She was slim and had high-set cheekbones and long silver hair. Her expression was always pleasant, and her eyes were kind. She always made me feel welcome. She would talk to me in the kitchen as she worked and on the porch from her rocking chair when she was resting. She brushed my hair every day on the steps of the porch, and those were also the times she used to teach me. To this day, there has never been anyone I've felt safer with or more loved by. She was my true protector growing up.

My grandparents fought every day, mostly about food. They weren't serious fights, just the kind people who've been married fifty years have. My grandmother always used chickens, roosters, and goats in her Santería rituals, and my grandfather hated it. It would've been fine, he said, if she earned enough money to cover the cost of the animals, but my grandmother wasn't the accountant my grandfather was. For her, it was more about helping the people who came to her, so she never charged them. My grandfather saw it as people taking advantage of her kindness. And he hated that.

When the rituals were over, my grandfather would collect the dead animals and take them to the market to sell. Or he cooked the meat to eat. "*¡No puedes vender esa carne, tiene mala energía!*" my grandmother always yelled. "You can't sell that meat! It has bad energy!" Still, he'd still collect the meat in a sack, picking up the pieces with the edge of his machete. She'd rush outside, and they'd fight over the bag. It was a real tug of war, and my grandfather almost always won. Despite all the things he saw my grandmother do, he never had any faith in Santería. Like a lot of Cuban men, he was an atheist. He would collect the meat after the ritual, eat the cookies and food left to the saints, and ignore all the warnings my grandmother gave him. Nothing ever happened to him. Maybe for Santería to work, you need to have at least some faith in it.

When I was seven, my life changed. Until then, I'd seen my grandmother's skill and power through the things she did for others, and little things she did for me. Mostly, she'd help me with a rash or a cold or some other minor infection—simple things, mostly. This time was different.

There was a lake by the house, and all my friends and family would go there to eat and relax. Half of the lake was full of freshwater, and the other half with seawater, where the frogs wouldn't go. The water would glow bluish-green there at night. One day, as my cousin and I were walking to the lake, two drunken men approached us from behind. My cousin saw them and ran away, but I didn't notice them until after she had started running. They started touching me and then grabbed me, but I wouldn't let them keep a hold on me. One of them punched me, and I fell into the lake, unconscious. I don't remember anything that happened after that. My cousin ran straight to my grandfather, who ran back to the lake and cut one of the men to pieces with his machete. The other one ran away. Grandfather didn't know either of the men. He just knew he had to save me. My cousin told me later that I had stopped breathing. My grandfather carried me to my grandmother. She immediately began to pray and did a healing ritual over me. Two days later, I woke up.

When I woke up, I had a bruise on my face and couldn't remember what had happened. When I asked questions, my cousin told me everything she'd seen. She was there for all of it.

Every Sunday after that day and for the rest of her life, my grandmother wore white. She had promised to wear white if my life was spared during the ritual she used to heal me. She kept that promise until the day she died.

My next memory is from when I was about ten. Shortly after I started school, I got a terrible case of chickenpox. It got worse after a week, and my grandfather took me via horse and buggy four hours to Santiago. The chickenpox vesicles coalesced and formed large sacks of fluid all over me. I remember feeling terrible with a bad cough, too. The doctors in Santiago put me in isolation, and every morning they stuck needles into the sacks of fluid all over me and scraped the small vesicles down to my skin. By the nighttime, they would be full of fluid again. It was painful, and after a couple of days, they started to use restraints to hold me down because I stopped cooperating. My mother and father were working in Santiago, but the doctors wouldn't let them see me because I was in quarantine.

The hospital room I was in was dark and underground. It was made of cinder blocks, and there weren't any windows. It felt more like a prison than a hospital. When another week went by without me getting any better, my grandmother made my grandfather bring her to see me. She decided to take matters into her own hands.

My grandmother left Guantánamo Bay with a pair of chickens, a sack of candles, and other herbs and supplies. I'm not sure how she got into the hospital, or even how she found out which room I was in, but she succeeded. When she entered my room, I was so happy to see her. She gestured for me to be quiet. I felt so weak and depressed, it seemed like I would never get better again. My grandmother made me lie down and drink a green liquid she brewed from herbs. She then began to pray and started a ritual. It was the only one she ever did with me awake, but it was something I will never forget. She placed candles on the floor around the bed and then cut the chickens' throats. I could always tell when her saint was inside her because she began to talk and gesture like a Black woman and speak in some African language I couldn't understand. The liquid she gave me also made me feel weird. I could see and hear and feel everything that was going on, but I couldn't move.

After my grandmother cut the chickens' throats, she poured some of their blood into a large silver goblet that was on the floor in the middle of the candles. Then she drank from the cup and began spitting the blood on me. She had also collected fresh chicken poop from one of the farms near Santiago, and she spread it all over my body. I fell asleep at that point and don't remember what else happened.

When I woke up the next morning, my grandmother was gone. The candles and goblet were gone, too. My skin was still covered in dry chicken poop, and the drops of chicken blood were still there, but also dry and mostly on my face and chest. When the doctors and nurses came in, they were shocked. They asked me what happened, but I said I didn't know. They wheeled me into another room where there was a shower and washed me from head to toe with lukewarm water. When all the dirt on my skin was gone, they were amazed to see all the chickenpox vesicles had dried up. They couldn't believe it! *I couldn't believe it!* This was the first time since the vesicles had appeared that I began to feel like myself again.

One Sunday in the summer, when the sun didn't set till late in the day, my uncle decided it was a good idea to fight with my aunt in front of everyone. He was a wealthy man (his family was from Spain), and he owned a fishing company. He had married my aunt when they were both young, and she was extremely beautiful. But she was also bipolar and had terrible mood swings. My uncle wouldn't hesitate to use his fists to "adjust" her mood. He would do it privately, and he would do it frequently, and my aunt would come to my grandparents' house every few weeks, bruised and hurt. After the first few times, she didn't bother to lie and claim it was an accident. Everyone knew what was going on. My uncles had repeatedly tried to confront her husband, even beating him up a few times, but my aunt always took him back and begged them not to hurt him. Sometimes she even attacked her brothers when they would come to help her. They finally got sick of her protecting him and decided not to get between them anymore.

This went on for years, and, of course, my grandmother knew what was happening. But she had never actually seen it happen. One day, my uncles and some of the local men were outside drinking, sitting in a semicircle next to a tree. My aunt walked up and asked her husband to come see the children dancing inside the house. He said no, and my aunt shook her head, said something under her breath, and walked away. There was something in the way she turned her head and walked away that flipped a switch inside him. He leaped out of his chair and ran after her. One of my uncles yelled at her to warn her, and she began to run toward the house. My uncles got up to run after him, and the local men sitting with them did too. My grandfather, carrying his machete, followed. But they were too far behind my aunt and her wife-beating husband.

She made it inside the house, but her husband was right behind her. He grabbed her by the shoulders and twisted her around, and then—in front of a room full of women, children, and my grandmother—he hit her in the face with his right hand. My grandmother immediately stood up, looked him in the face, stuck out her hand, and told him, "With the hand that you hit her with, you will never touch her again." Then she bowed her head and began praying. My grandfather walked into the house just at that moment. When he saw what had happened and heard her praying, his eyes opened wide. He bowed his head, turned around, and immediately walked out of the house. When my grandmother finished praying, my aunt's husband changed immediately. He said his right side was becoming numb, and eventually, he couldn't move his whole right side at all. They took him to the hospital. The doctor said he'd had a stroke because of the stress of the fight with my aunt. But everyone in the community knew what really happened. He had a stroke because of what my grandmother did.

A lot changed after that. My aunt stayed with him, but he never moved his right side again. He never spoke to my grandmother again, either. The people in town changed toward my grandmother after that, too. Before that happened, she had given out meat from her freezer to every person who would give her a sob story. She was just like that, and though my grandfather hated it, he finally came to accept that's just how she was. After the town heard and saw what happened to my aunt's husband, the word spread, and people treated my grandmother with a lot more respect. The sob stories stopped. They had known she was a healer, but they hadn't really known how powerful she was. After that, they knew, and they were different. They stopped asking for food and started giving her more money when she helped them, even though she still never charged.

People treated me differently after that incident, too. They knew there could be real consequences if they treated me otherwise, not that they were ever mean to me. They were just extra nice after that. My grandmother was good at what she did, healing and protecting people, but people saw another side of her now. She didn't get any busier because they already knew she was good, and she was already busy, but they would approach her with more respect. They stopped asking her for animals and things and began bringing her more gifts.

A Witch Finds Her Purpose

I grew up in Laredo, Texas, right on the border with Mexico. Back then, the border wasn't like it is today. It was much more open. Every day, there were at least half a dozen migrants walking through our yard from Mexico. I lived with my grandmother, who was a closet witch. She was a benevolent soul, always helping the people who came to her seeking help. Only our family and the people who came to her knew who she was. She did her work quietly, but I never forgot the things I saw her do.

My grandmother raised me until I was about ten years old, so she was the only real mother I knew. It wasn't until I left her home that I understood that what she was doing was not "normal." She had an altar, and I would watch her do spiritual cleansings with an egg and candle work where she would help her visitors understand their future. She also did other rituals and would never charge a dime for the work she did.

There was an episode I remember vividly. A beautiful young, pregnant migrant woman came to her house. She'd been raped and didn't want the baby. The word had spread among the migrants about what my grandmother could do. I could see it in the woman's face when she stood at the door. She knew my grandmother could help her. I don't know exactly what herbs my grandmother used. I just remember the woman drinking a tonic my grandmother made for her. A short time after she drank it, she began bleeding and crying in pain in a tone of voice I'll never forget. It was a shrieking cry, and I've never again heard a sound like it. There were buckets and buckets of blood. To this day, it was the amount of blood that I remember most clearly, even more than the crying. The woman endured the abortion and left our home the next morning as quietly and quickly as she'd come.

I lived with my parents after I turned ten, but I realized something was missing in my life when I was sixteen. I decided to ask my grandmother if she'd teach me how to be a witch. Her husband, my grandfather, was dying, and she was having a difficult time. She didn't have a lot of time for me, but she taught me enough to get me started on my path to becoming a witch.

I learned how to do cleansings with eggs and various herbs. I also learned candle work, binding work, and hexes. I also learned how to work with a pendulum. When I started with the pendulum, I had no money to buy a pendulum, so I made one out of my hair and an orange peel. That pendulum always worked. I would ask yes or no questions, and it would always tell me the right answer, no matter what I asked. As I grew older, I bought a tiger's eye pendulum. It always keeps me balanced between the spirit world and the real world.

I don't see my pendulum work as a portal to working with a spirit. I see it as a bridge to my real self, the more intuitive, confident side of my personality. I use the pendulum to make decisions and gain insight into my life. Over the years, I have asked the pendulum to guide me with jobs, help with my children, and many other things. I've never been led astray.

I married young and had children with a man who ended up being violent and dangerous. I didn't know he was that way when we first got together. Little by little, the real person underneath began to be revealed to me. He ended up going to jail but only received a short sentence, just four years. But my binding work has kept him in prison for much longer.

Binding work and freezing spells are common spells used to keep dangerous people and others we don't want in our life away from us. It is dangerous work and can have terrible consequences. You have to be sure it's necessary. Every witch does things a little differently. I do binding by writing the person's name on a small piece of paper, winding string around the paper, and tying three tiny knots as I repeat a specific phrase. Then I bury the paper. This has helped keep my violent ex-husband in jail and out of my life for many years, but I didn't realize there would still be some danger from other members of his family.

When my twins were only a few months old, I got a call from my ex that his cousin would be in town and wanted badly to see the twins. I'd never met her, and though I wanted nothing to do with my ex, I still let his family be involved. I didn't see much harm in letting her visit our children. The first time this cousin came, she showed up at my house around 8 p.m. She came, went into the bedroom to see the kids, and stayed for about an hour. We talked, and everything seemed fine. I thought this would just be a one-time thing.

Over the next couple of weeks, however, she came to my home every night and spent time with the twins and me. It seemed a little odd at first, but after a week or so, I got used to seeing her and started to enjoy the company and the help. Things seemed fine, but then she suddenly stopped coming, and I didn't hear from her anymore.

A few days after she stopped coming, I began to hear the sound of babies constantly crying in the house. They would be crying all the time, and my sister could hear it, too. We would run and check on the twins, but they would be fine. They were healthy and growing normally, and everything would always be fine when I checked on them. The crying grew so common I began to think I was hallucinating, but my sister heard it, too, so I knew it was real. I became depressed and exhausted. I couldn't sleep or concentrate because of how loud and persistent the crying became.

I went to the doctor and told him about the crying I was hearing. He blew it off and said I was just depressed and stressed from being overworked with the twins. He gave me some pills for depression and sent me on my way. I took the pills for a few weeks, but I didn't feel any better, and the crying continued. The pills made me feel sleepy during the day, so I stopped them after a couple of weeks.

Finally, I called my grandmother and told her what was happening. She had become so absorbed in caring for my grandfather that she couldn't help me, but she gave me the name of a woman she said would help. The woman was an older curandera. My grandmother said I could trust her.

I went to see the curandera, and when she saw me, she immediately knew something was wrong. "You're not crazy," she said before I said a word. "There is a woman who came into your home." She went on to describe my ex's cousin. "That woman had an abortion when she was seven months pregnant. She was carrying twins at the time. Those twins became restless spirits and have haunted her ever since she killed them. She sought you out not to see your twins, but to pass those spirits on to you. She needed someone with twins to do it, which is why you were chosen."

The curandera then stood up and passed six eggs over my body, one at a time. Then she took out some herbs and placed them in a Ziploc bag. "Burn these herbs on your stove when you get home," she said. Then she took a bag holding about a dozen limes and prayed over them. "Place these limes in the children's pillowcases and do not let that woman into your home again."

I went home, burned the herbs, and placed the limes in the children's pillowcases. The crying stopped immediately. That night, around 11 p.m., I heard a loud banging at the door. It was my husband's cousin banging on the door and crying, "You sent them back to me! You sent them back to me! I don't want them!" When I didn't open the door, she left. I called my father to come to my house—I was scared she'd come back and try something. Around 1 a.m., she did come back and started banging on the door again. My father went outside, told her to leave, and threatened to call the police. I've never seen her again.

I have since moved on and have a new husband. Things are good between us. He has a child from a previous relationship who is a troubled kid. He lived with his biological mother at first but would stay with us on the weekends. Things changed when we got a call from his mother one afternoon saying her son was no longer welcome in her home and that he would have to live with his father.

The boy was only ten, but he was already affiliating with gangs and getting into fights at school almost every day. The first day he came to stay with us, he was sent home from school after he shaved his eyebrows, a sign he had started affiliating with a particular gang. I'd gotten used to having him on the weekends, but I could tell that something had changed when he started living with us. He was angry all the time, and I became worried about the safety of my own kids. I did a binding spell on him, and things became manageable in the home after that, though I still felt uneasy.

A few weeks before the boy came to live with us, I had a dream. My stepson was running in a field in the dream, and San Miguel Caballero scooped him up while he was running. I thought it was a sign that maybe I should take him to learn how to ride horses, that perhaps that would help him get better. But it was much more serious than that.

A few weeks later, a coworker overheard me talking about the boy and some of his troubles. The coworker asked me if I wanted him to tell me more about the boy and my situation. He said he read tarot cards and that with his cards, he would help me understand what was really happening with the boy. I agreed.

We sat in the employee lounge, and he pulled out his phone and began looking at tarot cards on his phone. After a few moments, he looked at me and said, "The boy has been terribly abused in the home of his mother. Much worse than you know. It has become so bad that he feels great despair and wants to kill himself. He fights with everyone because he feels like no one cares about him, and everyone is his enemy." He looked back down at his phone, then looked up again. "The boy is a powerful medium and also a healer. It might be best if you showed him what you're able to do as a witch. It will help him to calm down and respect you. It will also help him on his path in the future." He looked down at his phone again. "You need to play with him and have fun with him. Whatever he wants to do. No one ever just plays with him. In time, the boy will come to view you as his true mother and closest and dearest friend." With those words, he looked at me, stood up, and said, "I hope that helps." He didn't ask for any money. His words touched me deeply, and everything he told me rang true.

I had told no one what had been happening in my house over the past few weeks. When the boy had run away from his mother's home, he had tried to run in front of a semi-truck. The driver said he saw a man on a horse carrying the boy and stopped just in time. When the driver said that, I knew it was my dream.

My stepson had been taken to a psychiatric hospital after his suicide attempt and told me and the doctors he was seeing an old man in our house. I thought he was hallucinating, but when my coworker told me the cards said he was a medium, it immediately made sense. I also suspected the abuse in his mother's home but didn't understand how serious it was. My stepson never admitted it, so we thought he was just angry or mentally ill.

A few days after my coworker did the tarot card reading for me, he pulled me aside and told me he had talked to one of his friends who was a shaman. The shaman confirmed that my stepson had the gift to become a medium and a shaman, but that at ten years old, he was too young to handle his talent. He told me to do a specific cleansing with basil and the branches from a couple of different bushes and to place seven glasses of water under the boy's bed while he slept. This ritual would take away some of his ability until he was old enough to handle it. I followed his instructions. The boy has improved.

People often pass judgment on me for being a witch. Many are uneducated or prejudiced and don't understand the things I do. I help many people, much in the same way I watched my grandmother serve many of the poor migrants who passed through her yard, people often with no one else to turn to. I'm not ashamed of the work I do. My work gives me contentment and peace, and I use my gifts to help my family and friends. I'm grateful for the skills I have and for what I learned growing up from my grandmother. I wouldn't trade it for anything.

A Santera Goes from Being Terrorized to Being a Healer

I met Linda on a mild and perfectly breezy summer day in Cuba. We sat by the beach and then in her home for several hours as she described her early childhood and path to becoming a santera. The first santera to agree to let me interview her at length, she encouraged me to continue my interviewing work—that it would be very significant for others in time. Raised in a devout Christian home, my initial preconceptions of mediums and the practice of spiritism created a very narrow understanding of the motives and paths many who participate in such practices experience. Linda's experience shattered those preconceived notions and forever changed my view of spiritism, and I came to understand the benevolent motives and intentions of those who often participate in such practices. Tormented as a child through no fault of her own, Linda went from being harassed and terrorized by unseen entities to cooperating with them to accomplish various works of healing and clairvoyance. Her experience illustrates how many individuals are often coerced into the practice of spiritism.*

**Contributor's name has been changed*

While most people become santeras only after going through a series of rituals, I never needed to do that. I was born with the gift in the way some santeras are. When I was seven, I began see-ing a small girl everywhere. She was dark-skinned, and at first, she would just stand and watch me. Later on, she began trying to talk to me. I was afraid at first because I was the only one who could see her. I would tell my parents about her, but when they looked and did not see anything, they would say to me she wasn't real.

In time, the little girl told me her name was Mama Francisca and that I was her special friend and that she was going to help me to help many people. What started as my just seeing her sitting in my house gradually became her being with me all the time.

She started telling me things about people. She said to me my father would have an automobile accident, and another man would die, but that my father would be okay. It happened a few days later. My father was driving a car, and another man died after he was struck. I had told my parents before the accident, but they dismissed it, and even after it happened, they still didn't believe me when I told them Mama Francisca had told me it would happen.

For years, I tried to ignore her and push her away. I would turn away and close my eyes when I saw her. But the more I tried to escape, the more she talked to me and tried to get me to talk to her. Nevertheless, I resisted for years. I didn't want to believe she was real. I knew I wasn't sick in my mind, but no one else could see her, and I couldn't explain what was happening. Eventually, I fell ill and began to feel sicker and sicker. I became weak and had trouble running and walking, and I got tired quickly. By the time I was eighteen, I was in a wheelchair. We tried going to doctors, but none of them could find anything wrong with me. I had already stopped trying to convince my family Mama Francisca was real. There was no point anymore. I was feeling run down all the time, and no one believed me. I was get-ting sicker and sicker.

Everything changed one day when I was in the post office with my mom, and an older Black woman walked in. She walked up to me and said, "You're going to be paralyzed soon. You need to start working with Mama Francisca, or she's going to keep punishing you. You have the gift. You have to use it."

My whole world changed! Not only did this woman know about Mama Francisca and was able to see her, but she also understood what was happening to me. From that day, that stranger became my godmother. She has the gift and is an extremely strong santera. Every santera has a godmother or godfather, a spiritual guide who is more experienced in Santería. They help the younger santeras and devotees to understand their gifts and teach them how to use their gifts to help others. She helped me understand the ways of the santera and what my true purpose is. She also helped me learn about the saints.

That day, I asked Mama Francisca what she wanted me to do. She instructed me to make a doll, a little Black girl, and to put it in my room. I made the doll from cloth and string and placed it in my room next to my bed. Mama Francisca also told me that it was time for me to begin helping others. I started to do that the next day. I never completed the usual rituals, though; I didn't have to. I didn't need a babalawo to help me identify my saint or any of the other rituals. I remained abstinent and wore white for a year. I also realized that the more people I helped, the better I felt both mentally and physically. The weakness left me, and I regained the normal use of my legs.

Not all santeras have the same gifts. Some are healers. Some have clairvoyance. Most are mediums and are able to channel spirits. Some are able to perform protection and cleansing rituals. I am blessed to have all the gifts, and, with time, my skills and talents have grown considerably.

When I was first starting, my primary gift was clairvoyance. Mama Francisca would tell me what was going to happen to people I already knew or would meet. Usually, within a day or two, what she told me would happen did happen. Later on, I began to listen to her and other saints as they spoke to me. They revealed various details of a person's life to me, and I would know a person's problems, or if someone was trying to harm them, for example, if they had *the bad eye* on them. I would also know if someone had consulted a palero against the person, and what ritual the palero had performed against them. In time, and with experience, I would be able to know why a person was seeking consultation as soon as I saw them. They wouldn't have to tell me anything. I could also read their emotional state, if they were happy or sad, and what kind of person they were.

Most often, people come to santeras seeking help with their health or the health of a loved one. Men also come looking to attract women, and women come looking for help with men. Many times, the problems have easy fixes, like a man having trouble obtaining an erection with a woman. I quickly learned many simple natural remedies for that and other common ailments. For a woman whose man is having erectile problems, she only need obtain nine small coconuts the size of a grapefruit or smaller. They must be coconuts that have fallen from the tree. She then must boil these in a large pot, and she must drink the liquid inside the coconuts. This resolves the problem. Her man will not have trouble after the coconut cure.

As my knowledge and skills have grown, so has my ability to heal more serious problems and perform more difficult rituals. Recently, a mother whose daughter was a young doctor came to see me because her daughter had developed an inoperable brain tumor and had been hospitalized. The doctors said there was nothing more they could do. The woman's daughter was unconscious and was told she had only weeks to live.

I consulted with the saints and then told the mother to bring an egg, cotton, and a small wooden box and to come with me to the hospital. I took the egg and began to move it around the young woman's head several times, then placed the egg on the cotton and set it in the small wooden box. I showed the mother how to repeat this procedure, and told her to do it daily until her daughter was well enough to leave the hospital. Once the daughter was at home, they were to leave the egg on the top of a hill without breaking it.

The mother followed my instructions, and her daughter was discharged a short time later. The doctors are still running tests and studies to confirm the results, but she has had no further difficulties.

Sometimes, however, the answer from the saints is *no*. When this happens, I cannot heal the person. It may be because the person is evil or the sickness is beyond healing. I've seen this happen a number of times. When it happens to someone I care about, it can be very painful. I grew very close to a family whose daughter was fourteen years old. The girl had an advanced stage of cancer, and her mother asked me repeatedly to save the child. The girl would come to my house and say, "Save me, Auntie, save me." The saints had already told me she was beyond saving, and there was nothing that could be done. This experience was especially difficult to endure. She was a kind girl, and the family was very good-hearted. But santeras have limitations. There are things that we cannot heal.

Another time that was difficult was when a santero traveled a long distance to find me after seeing my face in his dream. His saint had told him to find me. He also had advanced cancer. I consulted the saints, and they told me to tell him there was nothing that could be done. In his case, my role was just to help him accept the reality of what he was facing. He was a good man and had helped many people, but his time had come, and he needed to accept it.

When I was in my thirties, I developed a severe heart problem. I knew something was wrong when I could not see below my waist. I would also get tired very easily. The inside of my heart, one of the valves, was leaking badly. The doctors told me I would need surgery.

I traveled to a large hospital in Cuba to have the surgery, but I consulted the saints before I went. They told me that the surgery would be performed in the morning, but that the doctors would have to stop operating by noon, regardless of if they had fixed the problem or not. When I went to speak with the surgeon, I told him what the saints had told me. He scoffed and then said he did not accept what I was saying. I could see the ignorance in his eyes. I needed the surgery; I didn't have a choice. The saints told me I would be fine, so I trusted them.

Around the time I was to have the surgery, there were seven others in the hospital for the same condition. We all were to have surgery in the next few days. Before my surgery, the doctor gave me medicine to put me to sleep and put a tube down my throat. Despite the anesthesia, however, I knew everything that was happening during the surgery. The doctors took longer than expected, and as noon was approaching, I spoke to them. "You have to stop before noon!" The doctors stood back from the table. The surgeon did not want to stop. He wasn't finished. But an unconscious patient with a tube in her throat had just spoken to him. They could not explain what was happening. Nevertheless, they closed the wound and stopped working on me.

When I woke up, I was fine. My vision was back to normal, and I've never had any heart problems since. I was the only one of the eight patients who had that same surgery who survived. My faith grew tremendously from that experience. *(Author's note: I have tried to reach the anesthesiologist and surgeons attending this surgery but have not been successful in identifying them.)*

While some conditions can be cured, others cannot, and I frequently have to continue looking after a person to make sure they are safe and healthy. There was a little girl whose parents came to me to help her with seizures. She was born with fluid in her brain and would have many seizures a day. I spoke with the saints, and they told me to make a doll for the little girl and place it on one of the upper shelves on my altar. I made it as directed, and every time before the girl would have a seizure, the doll would shake and fall off the shelf. From that moment on, every time the doll shook, I would pray for the little girl, and she would not have a seizure. Her parents were very grateful for what I had done. "She has not had one seizure! Thank you, thank you, thank you so much!" Over time, I've had to pray less for this girl, but the doll rests on my altar still to this day.

One time, another santera came to see me and pay respects to my saints. She was a younger santera who was just beginning her journey, a clairvoyant and a medium. But she didn't possess the gift of healing. Even though she brought me a large basket of sweets as a present for my altar, I could read her spirit. I knew her gift for my altar was not sincere. In her heart, she was very jealous of me, jealous that she hadn't acquired the reputation or gifts I have. I told her to just leave the sweets, that I would take them to the altar later. I was going to just let it go, but she kept insisting she wanted to place the sweets on my altar herself. I finally said okay, but I did not give her permission in my heart. Her gift was tainted because her heart was not right.

If a santera does not permit you, you cannot enter the room of her altar. If you do, the saints will punish you. And though she was a santera, her fate would be no different from other people's. When she entered the room of my altar, she immediately fell. She broke her leg and began to cry out in pain. Because I'm too old to be lifting a grown woman off the floor, my children came into the room and helped her up. As they carried her out and back into the living room, I spoke to her. "You are a young santera," I said, "but you know enough to know you don't disrespect someone's altar. You were offering your gift with malice in your heart, and the saints punished you. Learn from this. Let it mold you and become a better santera. You have to keep your heart clean. There is no room for jealousy in a santera's heart." We called the paramedics. She was taken to the hospital and recovered without difficulty.

Bad things can happen from the actions of santeras. But those things are always the consequences of the actions of the person receiving them. I often see people who are the victims of deliberate malevolent rituals from others. When I cleanse a person, the consequence is often that the badness will return to the one who harmed the person. It is that individual's own actions that bring the badness back to them. I never feel guilty when a person intentionally harming someone reaps what they have sown.

I've become a very well-known santera in my community. Even people from other countries come to ask me to perform consultations with them. I cannot go anywhere near my home without people stopping to ask me for help. My phone rings constantly with requests. My family, who once ignored and discounted my accounts about Mama Francisca, are now my most common consultees. I do not charge for the work I do; no true santera does. Oftentimes, the people I help do not give me anything, whereas others can be extraordinarily generous to me, for which I am very grateful to the saints. Though the life of a santera is difficult, to be born with the gift is still a blessing in my eyes, and I am grateful for the work the saints have done through my life and hands.

While he was interviewing me, the writer asked me if I liked being a santera. It's not a life I would choose. It's a difficult life, but it's also very satisfying. It's something I have to do. There are many people, like the little girl with seizures, who require me to continue in the work I do. I must protect them and help them. It is a life of self-sacrifice, one that requires patience and discipline, but I'm grateful for what I am.

A Childhood Observer Recounts
Her Experience

I grew up in a city named Manzanillo, a coastal town about 200 kilometers (120 miles) from Santiago. There's a saying I heard a lot growing up: *Between husband and wife, no one can get in.* It's a tradition people in Cuba people still greatly respect. A man could smack his wife (or vice versa) in the middle of the street, and no one, not even the police, would intervene unless there were serious injuries. I grew up next to a couple that fought all the time. I would hear them almost every day. The husband liked to drink, and he was a mean-ass drunk. He would drink from morning to afternoon and then come home. And then the screams would come.

One day, the wife got sick of the beatings. The husband came home, drunk as usual, and went through his usual routine of beating her, but this time he was unusually cruel. The wife let out a scream that I've never forgotten. When the husband passed out, she finally got her revenge. She tied him up with rope and took a skillet to him, breaking both his arms and nearly bludgeoning him to death. That was the only night I ever heard him scream in pain. After that, I never heard him beat her again.

My grandmother, Nilda, and my grandfather never had physical altercations. My grandfather wasn't that kind of man, and he also knew Nilda wasn't the kind of woman that would put up with any beatings. She was a strong woman and had a great sense of humor. Sometimes we would sit on the balcony of her apartment, which over-looked the town square, and make fun of people as they walked by. She was fun, but she was also a vindictive and petty woman. If you wronged her, there would be no limit to what she was capable of doing to pay you back.

Although both of my grandmothers practiced Santería, they were very different. My mother's mom didn't consider herself a santera but saw herself more as a curandera and a medium. She would help people, often without pay, and the house would often be full of people participating in various rituals. My father's mother, Nilda, practiced a mixture of Santería and witchcraft in private and focused mostly on our family. No one ever came to her for help. She sometimes used her gifts to help me if I was sick or having problems, but she also used her gift or consulted with others stronger than herself to harm someone if she didn't like them or if they had wronged her.

My father was the only boy Nilda had, and she loved him in a way only a mother can. No woman was ever good enough for her son, and she saw to it that none of them stayed in his life for very long. She often used Santería to harm the women in my father's life, including my mother, too, because she never liked her.

Nilda did things all the time to try to poison my parents' relation-ship. Sometimes she would hire women to try to seduce my father. Though she tried to harm my mother with witchcraft, the curses never lasted very long. My grandmother did powerful protections for my mother and was able to know if people tried to harm her.

One time, my mother noticed she was habitually performing poorly at work, which had never been an issue for her. She was always exceptional at her job, so when she started having problems, she knew something was wrong. Her mother also knew something was wrong, so she used me in a ritual to find out exactly what someone was doing to my mother, who no longer participated in anything to do with Santería.

We went into the room with my grandmother's altar, and she began consulting with her spirits. Then she walked with me to a specific spot in an unused area in the city cemetery, where she started digging. To my surprise, she removed a small metal box from a small hole there. She opened the box and found one of my mother's books in it and some other small items. She put the box on the ground, opened it fully, and then peed on it. I was just eight or nine, and I wondered why my grandmother would pee on something, especially in an area where she might be seen. Later, she explained that peeing on the object would take away the work that Nilda had hired someone to do to my mother.

This was not the only time Nilda had done things like that to harm someone. Her actions taught me a valuable lesson. Because I grew up exposed to both Santería and brujería (witchcraft), I know that their powers are real and that there is definitely a difference. Nilda would both help and harm others, while my other grandmother would never do anything to harm anyone. Still, I respected them both and the powers they had because they both loved our family and did what they could to keep us safe.

The ability to be a santero or santera often runs in families. Nilda never either encouraged or discouraged me from pursuing Santería or brujería, but my other grandmother often encouraged me. I've seen spirits from the time I was a little girl. I know they are real. But I've never had an interest in becoming a santera. Even when I was young, the spirits would touch me at night and harass me by pulling the blankets off me or whispering in my ear while I was trying to sleep. When I was in my twenties, it got really bad, and one night I stood up and yelled at them, "Either kill me or leave me alone!" It got a lot better after that. Now they only occasionally try to talk to me and only rarely touch me at night.

I'm not sure if the power from spirits or Santería is from God. That's the main reason I never had a desire to work with the spirits. My mother started reading the Bible and going to different churches when I was young, and she wanted nothing to do with Santería after that. For her, it came from the devil. I imagine that for some, the temptation to become a santero is strong. To have access to gifts ordinary humans don't have is powerfully tempting. That's easy to understand. To be able to have some glimpse of your future life and the lives of others, to heal others, to protect your family, and to feel closer to something greater than yourself is a beautiful thing. But that's not something that has ever appealed to me. I've seen the lives of many santeros, and their life is a difficult one. It is a life of devotion and sacrifice. Though my grandmother never considered herself a santera, she was as devoted a healer as I've ever seen. Every morning, she took care of her saints. I'd always wake up and find her in her room praying with candles before she did anything else.

I've known many santeras and mediums. Married women seem by far to be the people who visit them most often. I've seen rituals with my grandmother and heard many stories about women wanting to know if their husbands are cheating or asking santeras to keep their husbands from cheating or mistreating them. I've never understood that. Men are who they are. I think back to that couple I saw when I was a child and remember how that wife handled her husband's abuse. That's the kind of woman I've always been. I've had many friends whose husbands have been unfaithful and abusive, and they just put up with it. I often say to myself, "He don't sleep? Tie his ass up when he's asleep, or make it seem like you wanna do something to spice up the sex. Once you've got him tied up, you can have your way with him. Put a knife to his throat or his balls. Choke him with a bag. Do whatever you want. Men are stupid. Once you show them you won't take their bullshit, they'll either leave or do right. You have that power as a woman."

I'm a close friend of a santera who knew me when I was young. She always tells me she never speeds things up or slows anything down when it comes to relationships. Everything has its time. If it's supposed to happen, it will happen. I can see the wisdom of that.

There are a lot of people who use Santería and santeros as a crutch. Every little problem, they run to get guidance or get their future read. They're the weak ones. Protection, healing, getting help with real problems, channeling spirits—that's what real santeros do.

In Cuba, there's a special place called El Cobre (The Copper). It's the most famous church in Cuba, and it's located in Santiago in the foothills of the Sierra Maestra Mountains. Over the years, thousands of people have gone to El Cobre to express their gratitude to the Virgen de la Caridad del Cobre for helping them. President Obama visited this church, National Security Council spokeswoman Bernadette Meehan reported, to "honor the sacrifices that Cuban Americans have made in their pursuit of liberty and opportunity, as well as their extraordinary contributions to our country." In so doing, the president was following the precedent of many Cubans, including some of the most beloved figures in Cuban history. There are countless stories of people who have journeyed to El Cobre to show their gratitude for the help they received from Ochún, la Virgen de la Caridad, Che Guevara and Fidel Castro among them. Many Cuban-born American athletes have returned to the church to express their gratitude and to leave gifts to show appreciation.

A Wayward Son Helped to Return Home

Clairvoyance is an extraordinary ability and one of the routine actualities of those who interact with unseen entities. It was this ability more than any other that convinced me of the existence of unseen entities. A difficult concept for those without experience interacting with such individuals, the account below is a typical representation of the clairvoyance those involved with various forms of spiritism manifest.

Several years ago, when I was barely a teenager, I came home from school in the middle of the day for lunch and found my father distraught. My twenty-one-year-old older brother, who was one test away from finishing his first year of nursing school, had run away to Havana in the middle of the night before his final exam. Two days later, my aunt called from Havana and told my parents that he was there. A month before he left, our family had gone to visit our aunt, and I had noticed that my brother and my cousin were spending a lot of time together. They had been close from the time they were young, and as I think back now, maybe a little too close.

It had been seven years since our last trip to Havana, during which my brother and my cousin had taken a romantic interest in each other. They were the same age, and they seemed to have an instant connection. But it was obviously more than family members should share. She is his first cousin, clearly not someone to cultivate a romantic relationship with. The twelve-hour train ride was enough to keep them away from each other for years, but when we went back this time, it seemed like nothing had changed between them from the time they were younger.

Havana is really different from Santiago. The shopping malls are better, the restaurants are better, the downtown is livelier, and there's much more nightlife. We were there for only a week, but we fit in as much as we could. We went shopping and to the clubs and the beach. My aunt and my three cousins went everywhere with us. Looking back now, I wasn't paying as much attention to my brother and our cousin as I should have. They were always walking and laughing together. I didn't imagine that they would even think about letting something romantic develop between them. But that's exactly what happened.

When my brother left home so suddenly, my parents and I were shocked. Even his friends knew he was making a terrible decision. He had only one test left to complete his first year of nursing school. In another year, he would finish school and be able to begin a career as a nurse. I couldn't understand why he would make such a foolish decision and give up his career when it is so hard in Cuba to have a stable life and make a living good enough to be able to provide for your family.

His decision was just so foolish. My parents couldn't understand it. It was so unlike him. He had never been irresponsible in that way. He'd never loved school, but to throw his chance at a career away to pursue a relationship with a family member? That's unthinkable to any reasonable person. It was not how he was raised. He knew his action would be entirely unacceptable to my parents.

There had to be more to the story. My parents went to see an espiritista to try to make sense of what was happening with my brother. They wanted to find out if there was anything that could be done to help him come home.

The truth of what was happening was even more shocking than we realized at first. We soon learned that my brother was living with my aunt and cousins. My aunt actually knew about the relationship and approved of it. More than that, she had been working for months to get my brother to move to Havana to help her. She has never had a steady job in her adult life. She sells anything she can to get by—drugs, sex, electronics, shoes, basically, anything she can get her hands on. One of my friends is an expert in computer security, and once we got into my brother's Facebook account, we saw months of daily messages from my aunt telling him about how great Havana was and how he should come to live with her there.

But that wasn't all she was doing. When my parents talked with the espiritista, she confirmed that my aunt had also been doing brujería on my brother. The espiritista did a ritual and dropped a coconut, which broke into several pieces with the shell side down. That was a very bad sign. My brother was to have many different problems, as we already knew, given the circumstances of his leaving. We also learned about many things he had been doing that he was hiding from our family. He was becoming an alcoholic and had been using drugs regularly for months before he left. He had stolen things from our home and our grandparents' home. All of this was so out of character for him that I knew someone must have been doing some brujería on him. The more my parents thought about what he was doing, the more they realized he needed to go through this. Despite how painful it would be, he needed to learn a lesson he would never forget.

The espiritista told us her conscience wouldn't allow her to let my brother stay in my aunt's home. She had known him from the time he was a young boy, and in many ways, she was like family to us. She said she was going to go to Havana herself to get him, but my father insisted that she not go. Then she asked me if she could do something to make our aunt's oldest daughter go crazy so my aunt would know what it's like to have a child taken from her. But I told her not to do anything against the family. There was also a young child in my aunt's home. One of my aunt's daughters had a six-month-old baby, and anything that would affect someone in the house could also affect the innocent baby.

It's a difficult thing to watch someone you love make bad decisions. It's even worse when you know members of your own family are working to bring evil to the family for their own selfish reasons. A curandero I know says that karma is a powerful thing and that the evil people do will always return to them. Though we know many powerful individuals who can make things happen through their work with spirits, we understand the repercussions of their work and how powerful it can be.

Eventually, my brother came home on his own. Maybe it was something the espiritista did. I don't know. Things were never the same in the family after what happened, but I'm grateful that he at least has a chance to choose a better path. Sometimes just having a curandera or espiritista provide insight into things that are hidden is an immense help.

A Counselor Describes His Experience as a Clairvoyant

I'm a medium. But my experience is not like what most of the mediums and curanderos I know describe. When I was young, I never saw spirits or heard the voices of spirits. No one in my family was a medium or associated with spiritism, the occult, or anything even closely related to spirits. I didn't realize that I was any different from anyone else until I was in my late teens.

Though I didn't fully appreciate it at first, I've had clairvoyant dreams since when I was a child. At first, the dreams were small and insignificant images of my life in the future. I would see myself in high school and then in college. Usually, some immense architecture images would be in the dream and define the location for me, and then I'd be doing something like laughing with friends or sitting in a lecture hall. The dreams were never what I'd consider important events, but, without fail, when they came true and I experienced what I'd seen in the dream, I knew that it was the real-life manifestation of what I'd seen months or years before.

I didn't know what to make of the dreams. There was no one I could talk to about them because I didn't want people to think I was crazy. And since the dreams never seemed important, I didn't give any significance to having an ability that never made any difference in my life or anyone else's. The dreaming just seemed like one of those life puzzles I had no desire to solve.

Things changed, however, when I was in my late teens. I began hearing the voices of people close to me in my head. The voices were not persistent or detailed or disturbing, usually just expressions of their emotions and feelings about something. Hearing them didn't irritate me, but I couldn't explain what was happening either. As it continued happening, I realized that I actually heard in my head the thoughts that other people were having. I pushed the thoughts away and just ignored them until something convinced me that I could hear other people's thoughts.

A colleague and I decided to take on a research project that would take about six weeks to complete. It was a difficult assignment. The work was tedious and detailed, and we met daily in the early evening to discuss the project and our progress toward its completion. About three weeks into the project, we were both starting to get fatigued, but I found the work rewarding and enjoyable, so I pushed through it. During one of our meetings, I heard my colleague say, "I'll be glad when we don't have to do this anymore." I looked up at him and said, "I will be too." The expression on his face was one of shock and surprise. That's when I realized that he had not said those words out loud, but had thought them. *I had heard in my mind what he was thinking.* Neither of us could believe what had just happened, and we never spoke about it again after.

The experience of hearing what other people think has recurred countless times with many different people since then. As far as I know, I cannot read minds at will, but I've never really tried. Usually, I hear when people have strong emotions directed at me or related to something meaningful in their lives. Again, I didn't feel the need to understand what was happening with me or test the limits of what I could do. It's an ability that helps me to understand people's true feelings, but I don't feel the desire to do anything with it. I've never tried to sit and see how much I can find out about a person by reading their thoughts. It just feels like an invasion of privacy.

In time, I became a counselor and therapist, and the ability to hear people's thoughts has helped me many times in my work. But I never spoke about what I could do with anyone until I met a medium a few years ago. I met a woman sitting on a bench by the side of a road I frequently travel. As I got close to her, I could feel her energy in a way I had never felt with anyone else before. It was like we were already friends and had a fondness for each other without actually being acquainted. It was not a physical attraction but a spiritual one. I could see in her face that she felt the same way.

I sat down next to her, and we talked for the next several hours. She told me she was an espiritista and that I was a medium and a powerful curandero, though I did not fully know the gift I had or how to use it. She told me the ability to hear thoughts is just an indication that I can do much more like channel and communicate with spirits. I felt a relief that is hard to put into words as I learned that the things I could do had an explanation. *A curandero?* I'd never considered that. I knew I wasn't mentally ill, but I couldn't explain what I was able to do either. When she explained things related to the gift I had in more detail, I felt relieved knowing many people who have similar skills could help me understand just what I am.

Through her, I've met many others with similar gifts. Still, despite their encouragement, I've chosen not to go any further in developing my abilities as a medium. I like where I am. My ability helps me as a counselor. I understand my clients better, and that's a tremendous asset for me in my work. I'm too afraid to let a spirit enter my body. I can often feel it when curanderos and others work on their altars concerning me or those I care about. I've come to have many genuinely benevolent friends who I would've never thought of being associated with before meeting the espiritista.

There are still many things about being a medium and connecting with spirits that frighten me deeply. I chose not to go any further in getting to know what I have, and that is my choice. The power of curanderos and mediums is real, but I know that those gifts come with a price. A curandero's life is difficult and requires tremendous self-sacrifice, and for me, having a normal life where I can look after myself has always been what is most important to me. I do not judge the individuals I've come to know who've made different choices. They have often helped me through difficulties and given me insight into decisions that confronted me, and I am grateful for the assistance they've provided. I don't understand why God gave some people the ability to be curanderos and mediums and communicate with spirits, but I know there are such people. I know most people will find it hard to believe what I'm saying without actually experiencing them first-hand, but these things are real.

Over time, I've gotten to know more about what is possible with my gift. I've stayed close to the espiritista I met, and she and others with more experience have helped me understand more about spirits and both the dangers and benefits of what it means to be a medium. Many have encouraged me to work with my abilities more fully, but no one has ever pushed me to become a curandero. It's just not a path I feel is right for me. The espiritista gave me some good advice a long time ago that I've never forgotten. She said, "This is not a life you can be in halfway. Either you decide to work with what you have, or you let it languish. It's a decision you have to make on your own. You have to be comfortable with who you are and what you want. The world of spirits is dangerous if you try to just play with it."

As I became more familiar with the community of curanderos, mediums, and others who work with spirits, my abilities have strengthened despite my reluctance to work with them. I can now sense more clearly when others who work with spirits are nearby and when my friends are working at their altars with their spirits on my behalf. The best way to describe what I feel is that it is a sort of hyper-developed intuition.

In time, I began to feel a particular attraction to oracle and tarot cards, initially thinking the cards were a less-involved way to use my gift. My initial thought that tarot cards were somehow less powerful was just another way my limited understanding of the spirit world manifested.

Tarot cards are among the most powerful forms of divination. The first time I picked up a pack, and every time since, I could feel powerful energy coming from the cards. I knew immediately that I would be able to use them properly, even though I had no instruction in doing so, and none of my friends worked with them. One of my curandera friends told me that I was just scratching the surface of what I could really do in reading the cards. But the cards are enough for me.

People are sometimes amazed at what I can tell them with the cards. There is a common saying among tarot card readers that the entire universe is contained in a deck of cards, and you can access any information you want to know. I've come to see the truth of this.

From the very first time I read tarot cards, I could feel the presence of a spirit guiding my reading. I started by reading my own life and those close to me. I've come to understand exactly what is possible with the cards. With the cards, I'm able to know anything in the present: where lost objects are, what is hidden in a particular situation, what the dangers are with a specific course of action. These are things I can access at any time. For example, when a friend of mine lost his wallet and asked me where it was, I could not only tell him where it was but what he was doing when he lost it and if he would find it again. If I want to know where someone or some object is or what secrets a person is hiding, I can easily find out.

When I read the cards, a powerful sort of intuition helps me understand the meanings hidden in them. The best way I can describe this is to imagine looking at an inkblot. Any number of objects may appear in an inkblot, depending on who's looking at it. The images interpreted in an inkblot are highly dependent on the person viewing it. When I read tarot cards, it's like looking at an inkblot and then feeling a force guiding me to see the most appropriate image to the person or situation being considered and then understanding the meaning hidden in the image. Sometimes it's the emotion displayed in the card or the emotion I feel viewing the card. Sometimes it's the negative space of an object in the card that reveals its message. Sometimes it's some other, smaller image hidden in a larger image of the card that is the essential part of a reading. Those with the gift to read tarot cards will immediately understand the message of the cards and how it relates to the answer being sought.

The future is not as set in stone as some believe. I don't believe in destiny. There are some things that the cards can tell you about the future that will definitely happen and other things that may happen, but in the end, a person's future is primarily affected by the choices they make. Often, the cards will only predict the likely outcome of a situation based on a person's current way of life and their choices. At times, there are surprises that I can see through the cards, like an upcoming pregnancy or a loss of a job or health problems. People are often very grateful for the direction and insight I give them.

I prefer to not use the cards very much in my own life or the lives of my immediate family or friends unless they persist in asking me for help. I prefer to live as close to a normal life as possible. I like not knowing what will happen and just living life in the present, enjoying my family and friends. In the end, all we really have is the now, and worrying too much about what may come only gives people anxiety.

A Lawyer Describes Her Work with the Unseen

Not all who routinely interact with unseen entities do so for the healing and assistance of others. Some do so with the explicit purpose of bringing harm to others and feel completely justified in doing so. Below is the account of one such person. It was her account that helped me appreciate that my own perspective on such individuals lacked nuance. As I came to understand the nature of her work and her approach to interacting with unseen entities, I realized that even those who practice things considered darker sometimes carry out actions some may consider as perfectly just.

If people knew about all the things I've done, they would hang me like they did the witches of Salem. I've seen spirits since I was very young, but I'm not a santera and haven't paid to become one. Everything I have is natural, and I've experienced these things for as long as I can remember. To this day, I don't have any saints. I only work with the dead.

When I was a child, spirits would call out to me, and I could feel their presence. But it wasn't until I was twelve years old that I started going with my cousin to different people known for working with spirits and saints. People known as santeros, paleros, espiritistas, and babalaos helped me understand my abilities. When my cousin went to do witchcraft with some of them, I went with her to understand what I was feeling and to learn how to make sense of it. I went to many municipalities and provinces to see people dedicated to these practices. They all told me that I was also mixed with Shango, a Santería saint. As I continued to grow up, I continued searching for more answers.

Were any members of your family practicing similar things?

Not at all. They did not practice anything related to Santería or witchcraft, but they did believe in the Virgin of Charity because she is the *patrón* of Cuba.

When you said you would see those people with your cousin, searching for answers, what exactly did you want to know?

I just wanted to know more. I didn't always listen to my parents because I knew they wouldn't have approved of what I was doing. I was rebellious in visiting those places and people. Many of the people I visited could see visions in glasses of water. Now I can do this, too. Many of the people I visited were able to read tarot cards or caracoles or use a coconut as a divination form. Sometimes I anticipated things that were going to happen. I would see them in dreams or just have a feeling about something, and it would always happen. I wanted to know what caused this, what was making me feel these things.

Can you narrate some of the things that confirmed or made you believe in spirits?

When I was twenty-four years old, I always felt this presence sitting on my bed. My mother used to tell me it was my guardian angel. When the spirit sat on the bed, I could see its shadow and feel coolness near me. But I did not understand what it was. One day I decided to visit a man who was from Haiti and worked pure witchcraft. I'm not going to lie. He was not a very good person. If you came to him asking to remove someone from your path (like killing them, making them sick, or something similar), it would happen in a matter of days or even hours. This Haitian man chose me as one of his goddaughters and told me that what I had, had to be worked with. But I was too afraid to be involved in any of that because then I would be able to foresee things that would happen to my family members. I refused to have that burden on me. Then he told me that the gift I had was going to make me sick until I agreed to work with it and that it was going to be worse and worse until I accepted the gift. I kept on refusing to work and deal with anything of that nature.

How does one get chosen by an altar, meaning, become a goddaughter/ son?

When you are born with what they call a gift, your abilities make you stand out. You're different from a normal person. An altar is able to sense this, and so when you go to visit people with the same ability for cleansing or changing a life, the saint or spirit chooses you to be under him. The other person has more experience and can guide you and becomes your godfather or godmother. Many people pay thousands of dollars to become a santero, but few people are born with the gift. After I met him, my godfather prepared me a Shango, a small idol for my altar, and anything I asked for or wanted, he would help me achieve.

Give me an example of something you wanted that your saint helped you with.

One of the things I always wanted was to be a lawyer, but the only available scholarship granted to me was civil construction. I was not fond of construction, so I dropped out of college and got married. I graduated from law school at age thirty-five, which my Shango granted me despite having to endure some terrible personal problems at the time. There were things happening in my life at the time that I cannot share. Even so, I received a law scholarship and graduated as a lawyer. That was one of the things that convinced me Shango was taking care of me.

After I met my husband, I started to seek God. I started going to church with him, though he was just my boyfriend at the time. Though not a santero, he also worked with spirits and saints. We wanted to get married in the church, so we gave our saints to the church, meaning we committed to not working with them again and only living by the Bible's rules. I got very sick when I stopped working with spirits. I was hooked to an IV and vomiting for six months, something that appeared to be white foam. I also kept seeing all kinds of weird things. I was diagnosed with a severe case of gastritis, which gave me unbearable pain.

I remember that when I was in my room on bed rest, I would see this short, black figure, like a small, dark man. The more pain I had, the more he laughed and kept staring at me and then laughing some more. The pain was so severe that I told my dad if it didn't go away soon, I was going to burn myself alive.

When I'd had enough of the pain, I went to the house of a man who lives outside the city. He works with spirits and saints and also the dead. He said, "Your world is not of light, but of pure darkness. Do not look to the church because you do not belong in the church." Well, I said to myself, that might be true. Soon, I gave up on the church and all the hypocrites that called themselves Christians. I had tried my best to find God and given up working with the spirits, but nothing good came from it.

That same day, the man took away my illness with a simple cleansing, and I resigned myself to going back to my spirits. He prepared them for me, and since then, I have not had any more diseases.

When I moved in with my boyfriend, I would see his dead grandmother sitting in a rocking chair in the living room. When I woke up in the middle of the night, I would also see her then. I described the woman to my boyfriend and told him exactly what she looked like, a person I'd never seen in my life. It was his grandmother exactly.

My boyfriend and I got married by the court, and, little by little, we started building his altar. He's always been one of the chosen ones gifted with clarity and vision. He can speak to saints and has a light that allows him to heal and work with children and pregnant women. He's the opposite of me because I can only work with the dark.

Everything I do comes from darkness, but, like a business, when you take, you have to give something in return. As the old saying goes, *You give to me, and I give back to you in return.* When I work with a spirit that appears as an old African woman, I have to make a sacrifice in return for her help. She asks to be paid in blood. She used to accept either animal or human blood (only my blood), but over time, she has become more accustomed to human blood (still only my blood), and now that's all she asks for. I have her idol next to my husband's altar, surrounded by containers where I put my offerings. Every Friday, I smoke a cigar for her, drink a few shots of rum for her, and sometimes give her food and a red rose. I do simple things like that to show her she is being attended to. Whatever I ask her for, she gives me. She takes care of me and protects me. She is always by my side, and I try to do the same for her. I have seen this spirit from a very young age, and my husband is the only other person who has seen her. She does not let herself be seen by other people, even those who work with the spiritual world. She tells me that she was abused by men when she was alive, and now she likes to punish evil people severely. She always dresses in black.

Today, I work with many people with my spirits, not as a business, but only with people I want to work with—my friends and acquaintances, not strangers. I have helped many friends and don't charge anything. They just bring me what I need: rum, cigars, and such; I take care of the rest.

How did you develop the ability to read in a glass of water?

I can see anything that is a spirit, the dead, or dark things in a glass of water. I am not clairvoyant, nor can I see the future, but I can see when a person's life goes away. I sometimes have to be around sick people for work, but I don't like going to the critical-state room. I know who is going to die because I see a light above their body. I can even see their spirit floating above their human form.

Can you tell me about your relationship with your husband because you're from opposite worlds?

We were like dogs and cats, but after ten years of marriage, we now have tranquility between us. There used to be a lot of friction between us because my dead spirits didn't want him near me, and his spirits and saints did not accept my practices. We went to see people with more experience with spirits than we have. They helped us reach an agreement, and now we understand that each person should be in their place. Respecting each other's beliefs and practices keeps us at peace. At the same time, we help each other. If I need his spirituality, he helps me, and if he needs my darkness, I help him.

Let me tell you a story about when we had to come together as one. My grandson is seven years old, and when he was born, my husband prepared an Elegua ritual for him as a form of protection. My grandson surprises me because he knows a lot about the spiritual world at his young age, things that have never been explained to him, such as the importance of attending to a spirit and the things that must be done to assist them.

Not long ago, the boy fell down, and they had to perform emergency surgery on him. We prepared a necklace with protections from the dead and saints to protect him, but his necklace burst the day before the operation. That meant something was not going to go well with the surgery. There was going to be a bad complication. We understood the sign and knew what we had to do.

My husband and I worked together to bring our grandson back to health, and I redirected the bad that was meant for him. Another child who entered the operating room died. It was the evil that was meant for my grandson. It was not that we wanted another child to die, but we can only dominate things to a point, and the bad always has to fall on someone else as an exchange. Let me be clear: it is not that doing this made me proud, but if bad things are meant to happen to you or your loved ones, and you have the power to change things, would you not do so?

Your husband likes to work with health and good things, and you can do the same. What gives you more enjoyment—good or bad?

Look, I don't mess with anyone, and I don't like judging anyone. I am very observant. I'm a woman of few words. But right now, I like evil more than good. People come to me to do evil things to others, but if the person they seek to harm has not done anything wrong, I do not mess with them. I refuse to do this type of work for money or any other reason. If it is a personal case of someone who is hurting people I care about or me, that person is fucked! It is not about always doing evil. Many people deserve to be cut in half. The correct word is *justice*. That is very valuable to me. I see it as giving people what they deserve. I don't hurt anyone who doesn't deserve it.

I don't like spiritual masses or rituals. I don't like *bembé* or people who sing or pray. I have gone to a few ceremonies, but only those that invoke the dead. There have been only a few times when I've gone with people to such things, but only to observe. When I have questions, I only talk to my godfather.

Years ago, I saw an image of the sun, the moon, and a star in a dream. The spirits told me that I had to get a tattoo of this image on the back of my neck. I didn't know what the symbol meant, but I got the tattoo anyway. I have never opened a book to learn its meaning. I believe in what I have, and the spirits make me understand everything in their time.

One afternoon I was sitting in my living room and started to see many symbols on the floor. I didn't know what they meant. I began to draw them, and then I went to a person with more experience. They confirmed what I already knew, that these symbols were demonic. This was a message to tell me that I had the power to do darker things, things much stronger. But if I start developing my power in that direction, I will be transitioning into a completely darker world and will eventually become a demon. I'm not afraid of any of that anymore, but I know that everything has a limit, and I feel good in the place where I am with what I'm able to do. For me, the meaning of dark is bad, but I also know that there's always good in bad and bad in good.

A Curse with a Pair of Golden Shoes

When you understand Santería, you realize everything is about energy. The things that shamans and others who can access the spirit realm accomplish are because spirits understand energy. The spirits teach these individuals how to change people's energy for good or bad, depending on how they use their gift. But there is a price for this gift and for the knowledge and abilities that come with it.

Santería and other similar practices can be dangerous if you're not careful about with whom you associate. Some gifted individuals manipulate people for money. At times they'll tell a person that they, the consultee, have power. They'll do this to connect with the person and get them to start paying them to make progress in their abilities or to help them when they have problems. Sometimes they will even make problems for the person to receive more money from them when they seek help.

I know that spirits are real; I've been seeing them since I was a child. When I was young, I would watch my family and friends celebrate *bembé* and other rituals. At times, even as a child, I would be included in the traditions, but my family only used their gifts to heal and help people. Though I had the gift, I never wanted to become a santera. I had a child when I was a teenager, and I didn't want anything to take my time away from her. I also married when I was young, only eighteen, and, unfortunately, I did not choose a husband wisely.

I married my child's father because I thought it was the right thing to do, and because I loved him at the time. He was everything a young woman could want. He was handsome, he knew how to talk to women, and he had money. He could push all my buttons, but in time, the magic wore off. We had a second child a few years after the first, and then things really changed. He started cheating on me a few years after we were married, and in time, I started doing the same to him. Our marriage became just an empty shell. The love was gone, but we stayed together for the children.

His mother and I never got along. She was a miserable woman, angry all the time. She was a Santería devotee, and she made it known more than once that she wanted me out of his life. One day, we went to spend a few days with her so she could have some time with the kids. I've always loved to go out and have a good time, and since we would have a babysitter and be in the city for a few days, I made sure I had enough outfits to make the most of our nights.

I've never stopped loving dressing up and wearing a beautiful pair of heels. At the time, I was working for an agency to secure contracts with various companies. I was part accountant and part saleswoman. I was good at my job, and financially, we were doing well. My closet was full of clothes and heels, but I had a couple pairs that I couldn't live without. My favorite was a gold pair with three-and-a-half-inch heels. They had laces that wrapped around my ankles and lower legs and made my legs look super sexy. When I wore them, I felt like I could conquer the world.

We went out a few times that weekend, but I decided to wear different heels because the gold ones didn't work well with the dress I finally decided on. I threw the gold heels back in the suitcase and pushed it to the back of the closet in the room where we slept.

The trip was uneventful. We spent a few days in the city and then came home. When I got home, I unpacked the suitcase. My favorite shoes were gone! I knew I had put them in the suitcase. Someone had to have taken them out. I asked my husband about them, but he didn't know where they were. I knew that it had to be my mother-in-law. I called her, and of course, she denied it. Our relationship was already bad, and I wasn't going to make an issue over a pair of shoes, so I let it go and moved on with things.

Over the next few weeks, I started feeling off. I didn't feel like myself. I was down a lot and moody for no reason at all. I started having problems at work, and we lost several contracts. Over the next six months, I didn't close any deals, I realized someone must have done something to me. I needed to see a santera.

A few weeks later, I went to see a santera my family knew. It had been years since I had seen a santera, but I trusted her because of the things I had heard my family say. I knew if she was as good as they said, she would be able to tell me if something was wrong.

She invited me into her home and had me sit down in a chair near her altar while she sat in a position that allowed her to look at her altar and me without moving. When she shook my hand, my wedding rings dropped to the floor. She looked at me and said, "Don't worry. You will marry again." She again looked at the altar and back at me and said, "Someone wants you dead. The only reason you're not dead is because of what your family did for you when you were young. Your grandmother is still protecting you. She's been with you as things have been difficult for you lately."

The santera was right. Though I couldn't put it into words, I had been feeling my grandmother's presence for the past several months. When I was young and near her, I always felt positivity close to her. "She's standing next to you right now," the santera said. "What they've done to you will take several sessions to remove. You have a pair of shoes that have gone missing?"

"Yes," I said.

"They're gold, right?"

"Yes," I replied. *I knew it,* I said to myself.

"Your husband's mother stole the shoes and took them to a bruja," the santera said. "She paid the bruja a lot of money. She wants you out of his life."

The santera then went into another room and got some large leaves, a glass of water, and an egg. She asked me to stand up, then took the leaves and began waving them in front of me. As she did so, the leaves withered and turned brown. She then passed the egg all over my body. When she broke it, it was black and hairy inside. I knew when I saw the egg that things were bad. I had seen that before, when I was young. It's always a bad sign.

She then took a bottle of rum and poured some on the floor in the shape of a cross. She lit the rum and made me walk around it, and as I did, a vortex formed about the height of my waist and followed me as I walked around the flame. When this finished, she had me sit back down and told me that was all she could do for me that day. She told me I would need to come back in a few days so she could finish cleansing me.

I felt much better after I left her house. It was like a weight had been lifted off me, but I still didn't feel like my old self. I went back a few days later to see the santera again. She started the same way on the second visit. She took the glass of water and set it down in front of me, then got three large leaves and began waving them in front of me again. But this time, the leaves did not wither, but only wilted slightly. She again lit the fire and had me walk around it, but this time, it stayed mostly in the center as I walked around it and didn't follow me as much. "The person who did this to you," she told me, "will receive what was meant for you. But you need to be very careful what you leave around your mother-in-law. She wants only bad for you." I thanked the santera for her help and left. She never asked me for any money, and my family had told me not to offer her any, that she would be offended if I did.

The next day I felt normal again. Things at work went exceptionally well for the next few months. I closed more than fifteen contracts, and the company could buy the building where our office was located.

My husband decided to leave me a few months after that, but it didn't hurt me. I felt more relieved than anything else. I had been hoping for years that he would leave, and when he finally did, I felt like I could finally get started on the next chapter of my life. I found out through the grapevine about the woman my mother-in-law had paid to try to hurt me. What the santera had said came true: the bruja had abruptly stopped working and was no longer taking clients. Within six months of my cleansing, both of her daughters died tragically.

I've always been mindful of how I treat people. Growing up around Santería taught me that. I don't need bad energy following me in my life. You don't know what people have protecting them, and when you do bad things, evil will come back to you. It's a lesson some people have to learn the hard way.

A Cursed Doll and an Untimely Death

You have to be very careful about what objects you bring into your home. I learned this after seeing what happened to a friend. She had a hobby of collecting dolls and loved to go to garage sales. Her apartment was full of dolls, and when she moved into a larger apartment after getting a raise at work, she decided to expand her collection significantly.

One day, shortly after she moved into her new apartment, she invited me over to see how she had decorated her new place. It looked beautiful. She had good taste, and there were dolls all around the apartment. We started talking, and she went into her bedroom and brought out a few things she had recently bought at a yard sale. Among these was a small back doll wearing white clothes. When she held the doll near me, it made me feel awful. The energy around it felt very dark.

"Where did you get that?" I asked her.

"I found it at a garage sale. It's cute, huh?" she replied.

"You have to get rid of that immediately," I said. "It has very bad energy."

"Come on," she protested. "It's just a doll. There's nothing wrong with keeping it. I don't believe in any of that stuff, anyway."

I left the house shortly after that and had a bad feeling all day after being near that doll. That night, I spoke with one of my cousins and asked her if I needed to do anything to get the bad energy off me. I wasn't sure if I was in any danger, but I knew my friend's keeping that doll was a bad mistake, and I didn't want any of that energy affecting me.

My cousin told me to do a cleansing and that I would be fine. I did the cleansing and felt better. I never went back to my friend's apartment again. We lost contact, and about a year later, I saw on Facebook that she had died from an aggressive form of ovarian cancer. She was only in her twenties.

A Medical Doctor Sees the Power of an Energy Cleansing

I was selected in 2014 by my province to fulfill a medical mission in Venezuela. It's common for Cuban physicians to spend a year or more serving in Venezuela's medical system. The pay was a little more than what I usually make in Cuba, and I thought the experience would be worthwhile, as I'd never left the island before.

I specialize in sports medicine and usually work with Olympic-level athletes, but in Venezuela, I worked with various patients, including pregnant women, children, and college-level athletes. I also spent time with many Venezuelan Olympic athletes. I have many therapeutic strategies and pride myself on helping athletes get the most out of their bodies and keeping them healthy to stay active in the sports they love. I've worked with gymnasts, dancers, and many Olympic martial artists, including karate and judo Olympians.

After I'd been in Venezuela for six months, another doctor from Cuba came to work with me. (The director of medical missions came with her and introduced us.) The new doctor was an attractive, confident, young woman from Santiago de Cuba, the cradle of Santería. She was also a sports medicine physician, and we started to work together in the clinics and training facilities. Because she was a skilled physician, we got along well. But after two months of working with her, I began to notice the smell of tobacco coming from her room at night. I never imagined what she was doing.

During my first six months in Venezuela, I also developed a close friendship with a Venezuelan physician I worked with. A few months after the new Cuban physician arrived, this man became very ill from a chikungunya infection. After a few days of high fevers, he died. It was a challenging time for me. We all needed a break, and so the director of medical missions suspended our clinic duties for a week.

A few days after the physician's death, I was eating in my room around nine o'clock in the evening when I noticed the young Cuban physician staring at me from her room. She had a very different look about her, different from every other time I had seen her. Her eyes looked different, her facial expression was stern and serious, and when she moved, her movements were deliberate and slow. She sauntered to my room slowly, and when she reached my door, she turned and began to point with her index finger toward her room. I stood up to speak to her, but I fell to the floor when I saw the inside of her room. I was terrified by what I saw: an image of my dead friend was standing right there. I lay on the floor, and when I asked the young woman what had happened, she started to have convulsions throughout her whole body. Then she started screaming.

I didn't know what to do. When I called the director to ask for help and told him what was happening, he told me there was nothing he could do to help because where I lived was a "red zone," a designation indicating a very high degree of danger and violence. He could not travel there at night. He said I would have to take her to the hospital. Because there were no ambulances, he added, I would have to take her myself.

But I couldn't move her. I was afraid, and she wouldn't cooperate. That night, I witnessed her repeatedly changing her personality and emotional state from laughing to crying and acting violently. She broke things in her room and around the house where we were staying. Her voice turned into different tones. I felt like I was living a nightmare as I watched her fight to regain control of her mind.

I stayed awake all night and kept watch over her. Around six o'clock in the morning, an ambulance arrived with some other Cuban doctors, and when they entered the house, I felt a great sense of relief. They had to sedate my colleague, and when she was only slightly calmer, they transported both of us to the hospital, where a Cuban ER physician treated her for four hours. They gave her repeated high doses of medicine that would typically put a person to sleep for many hours, but no matter what medicines they tried, her symptoms remained uncontrollable.

I decided to stay in the waiting room while the doctors worked. I was frightened by what I'd seen and didn't want to be near her. When she told the doctors she wanted to see me, I reluctantly agreed. When I walked into her room, she asked me in a voice that was not her usual voice to take her to the room where my friend had recently died. The ER physician reluctantly let us go to the ICU, but he followed closely behind us to observe. When we entered the room, an ICU nurse looked intensely at her and told us immediately that a spirit had possessed my colleague. The nurse said that there was nothing the doctors could do to help us and that we needed to look for a person who knew of spirits.

I left my colleague in the hospital and drove to another Cuban physician area. After three hours of asking various doctors, I spoke to an ER doctor who told me about a sixty-five-year-old man who could help us. He also told me where to find him. I went to the older man's home and spoke with him about the situation. He asked me to bring him some tobacco and branches from a tree, after which I took him to the hospital.

When we arrived and went into her room, the man took out the tobacco and the branches. As he began to smoke and speak loudly and strangely, my colleague's face began to calm and regain the color she had lost while experiencing her problem. When the man finished praying and smoking, she fell asleep. She stayed in the hospital for two more days and slept the whole time. I went back to work in the clinic, but when she awoke on the third day, the ER doctor called me and asked me to come to the hospital because she had asked for me. When I entered the room, she broke into tears and said that when she'd seen the image of my other friend who had just died, the pain was so great she'd lost control of herself.

We left the hospital and told the director we would never go back to that house again except to get our belongings. My colleague was told she had to go back to Cuba. They permitted me to stay in Venezuela but transferred me to a different clinic in another city.

I can't explain scientifically what happened. I don't know. Medically, no explanation fits. I've heard many stories of similar things like this happening to people who live in Cuba. It was easy for me to dismiss such incidents before I experienced what happened to my colleague. After what I saw and experienced with my own eyes, however, I know spirits are real.

A Masseur Brings Healing through Touch

People give me different titles—espiritista, witch doctor, shaman, curandero, healer. But the labels don't matter much to me. Above all, I'm a healer. Like many healers, I started when I was very young. When I was six years old, my mother went to see a curandera to do a cleansing. I'm not sure exactly what this woman was, but she had an altar in her home, and I felt immediately drawn to it. Because I was so young, I didn't sit in the waiting room with the other people waiting to see her while she did her work with my mother. Instead, I watched. I felt an overwhelming sense of pleasantness and peace in her home and also in the house of another curandera my mom would sometimes visit. That sense of peace was enough for me to want to know more about magic and the occult.

I've always been drawn to altars. The altar in the curandera's house reminded me of being in church. My mother was Catholic, and I've always been especially fond of the Virgin Mary. I would go to church with my mother and often spend time staring at the altars there and feeling a connection with them.

I grew up in El Paso, Texas, and a short time after I went to the curandera's house with my mom, I went to a store in downtown El Paso where books and other items associated with the occult were sold. That store had an interesting mix of things for sale. There was an enormous collection of costumes on the first floor, but to get to the store section that sold books on the occult, I had to pass through the adult section of the store that had porn and all kinds of sex toys. The occult section was not a place where kids were allowed to go, but I somehow made it through and was able to find a book called *White Magic*.

I was only six, but I bought it and read it from cover to cover. I became fascinated with the herbs and teas described in the book and was also intrigued by tarot cards. There were no tarot cards in that store, but a few days after I read the book, I bought a deck of Spanish playing cards because the book said they could be used in place of tarot cards. Most people don't realize how similar ordinary playing cards are to tarot cards. A deck of playing cards only has four fewer cards than the Minor Arcana cards in the tarot deck. They both have four suits and face cards, the only difference being that ordinary playing cards lack the page cards for the four suits, which accounts for the difference.

Shortly after reading the book on white magic, I found a cat that had been hit by a car near my home. I brought it home, set its broken bones, and nursed it back to health with the herbs and teas I'd learned about in the book. After that, I began reading tarot cards for a neighbor. At first, she thought I was just a cute little boy playing at the occult, but it didn't take long for my neighbor and my mom to realize I was not just playing with the cards. The occasional readings I did for my neighbor took place more and more often, and soon, my neighbor was inviting people over to have me do card readings with them, too. Everything was light and fun for the people who came.

But things changed when the cards predicted that my neighbor's mother would die. At first, my mom and my neighbor thought the readings were not serious. After we didn't see my neighbor for the next several weeks, I had a dream that foretold death, and in it, I was told by a human figure. The details, I cannot remember, only that my neighbor's mom was already dead. This was especially sad because she was still young and had just retired. She had started coughing up blood a few weeks after she retired and was diagnosed with stomach cancer a few weeks after that.

My mom started to believe in what I was doing. But things really changed for her when I had another dream a few weeks later. It was a dream that convinced my mother that my abilities were genuine.

In my dream, I was walking near a Mexican market in Juarez. An elderly, blind woman approached me and touched my shoulder and told me that my mother needed to avoid eating any pork for the next thirty days, or she would become very sick. When I woke up, I knew that the dream had special significance. It was so vivid that I knew it was not just an ordinary dream. When I told my mother about my dream, and to avoid eating pork, she dismissed it. A short time later, she ate some old pork rinds she found in a bag. She noticed that they didn't taste right, but she ate them anyway. After she developed a severe stomach infection, she remembered my warning and knew that I must have some real ability.

When I was around ten or eleven, I decided to make a Ouija board. I wasn't sure what would happen, but I tried to summon Beelzebub while alone with the board. I didn't receive any communication from the board, but I soon began to feel a powerful, heavy presence in my home, and so did my mother. Then our home began to have an unusually high number of flies. A few days later, I was walking up the stairs with a bag of chips and a bottle of soda in my hands. I saw a spirit at the top of the stairs. It was coming down as I was walking up. For some reason, I wasn't afraid, but as we approached each other, I felt the spirit pass through me, then nudge me. I lost my balance and fell down the stairs. That's when I started to feel afraid.

A few days later, my neighbor found several Catholic prayer cards in her house and thought maybe I might want them. She brought them over, and I studied each one. Among the prayer cards was the exorcism prayer of St. Michael. After I repeated this prayer around the house several times, the heavy presence I felt left the house, as did the flies.

Shortly after this incident, my mother became afraid of having anything associated with the occult in the house. She took my cards, the Ouija board, and the things I had for my altar and threw everything out of the house. It wasn't until I overheard a phone conversation that I understood what she was feeling. I heard her say to my neighbor, "He is worshipping Satan!" I felt offended and indignant. I said, "Mom, where do you see an image of Satan? I'm not worshipping Satan."

When I was in the sixth grade, I became a voracious reader. I spent hours reading encyclopedias and was deeply moved by studying religions and the positivity they can engender. As I grew in my understanding of God, the idea of something coming from nothing gave me no foundation for believing in God. The idea that God always existed and then decided to share the gift of life and wanted someone to love him and worship him never made any sense. I had believed this when I was a young child because that was what I was taught, but the story never made any sense to me. God was alone and then decided to make someone to love him? I thought to myself, *God needs a psychologist.*

As I continued my studies and research into religions, I became particularly fond of Buddha and Buddhism's history. I've come to see the divine in nearly all religions. To me, it's like the different flavors of ice cream: there are many flavors, and everyone has their favorite. When I found Buddhism, it just felt like home to me. Many religions are all about hierarchy—who is the greatest, who has the most knowledge, the most prominence. These thoughts often lead to patterns of behavior that lead to suffering. But Buddhism is about service, about purging the thoughts and habits that lead to suffering. The Buddha's practices and teachings augmented my commitment to healing and have aided me beyond measure in my path to bringing joy and health to others.

The more militant and strict a person is, the less they will experience spirits and magic. Real magic and healing are not in the cards or candles; they're found in meditation. True magic and healing are found in quieting our negative thoughts, in deliberative contemplative meditation that is not focused on clearing out one's mind entirely but on removing the toxic thoughts and habits that lead to our own suffering and the suffering of others. That's the real reason I became a healer. A genuine healer relieves the suffering of others. He doesn't profit from suffering the way Western medicine does. He heals himself first, and then he is free to guide others to health.

When I was young, before my first experience with the occult, I had a great love for the Virgin Mary and altars. I attended a Catholic daycare and preschool and remember asking the nuns if I could go outside and pick flowers to place in front of the Virgin Mary's large statue that stood inside the school. The nuns found it odd, but it was natural to me. To this day, I love the Virgin Mary and view her as being at the same level as Jesus, though I still consider myself more of a Buddhist than a Catholic. I keep dozens of statues in my home, statues from many different faiths and religions. The figures are mostly for the people who visit me for consultations to have focal points to direct their attention. I don't go into transition often. I don't have to. I speak directly to the spirits the same way I talk to people. Sometimes I'm directed through conversation. Other times, I'm directed through intuition or signs. Some Native Americans believe every animal you see crossing your path is a sign, but I've come to know signs can appear in nearly everything.

I've come to know what works for me in my religious beliefs and my healing practices and training. I've studied many different religions and paths for magic and communing with the spirits and healing. I run a small massage clinic out of my home, where I do healings and tarot card readings. The three main spirits I work with are Jesus, St. Michael, and the Virgin Mary.

Was there a particular experience with the Catholic Church that affected your perspective?

There was a woman who had a teenage daughter. She didn't want her to get pregnant, so she prayed every day and lit candles to the Virgin Mary. But the girl eventually became pregnant. Did Mary fail her? People often think of saints and deities as magic erasers. They believe the saints can undo any mistakes we make. But they do not, nor should they fix or correct our choices. We are responsible for our own decisions and the consequences they bring. If you want to avoid disappointment, never expect anything. It is we who set ourselves up for heartache. Disappointment doesn't come from the devil but is our frailty. This concept really appeals to me.

What healing techniques do you use in your healing practice?

I'm a licensed masseur. I have many clients who come for massage, but I can offer many other healing modalities. That is not to say massage is not powerful. While powerful results can often be achieved with simply massage alone, I often incorporate desempacho and acupressure into my massage techniques. I also believe in La Sobada. I've seen the reality of its principles. Any disease can be improved, if not wholly resolved, by understanding nutrition, absorption, and excretion. There are also many principles from Chinese medicine I find helpful when assisting clients.

Do you have any specific cases of individuals with complicated health problems you've been able to help treat successfully?

I take great joy in working with all my patients, but I have a particular fondness for working with women with fertility problems. To date, twenty-two women have come to me with difficulties with fertility, and I have helped all twenty-two to have a child. This is a great source of satisfaction and joy for me. Usually, directed massage is all that is needed to restore proper balance and blood flow to the uterus to relieve infertility problems.

One memorable patient was a woman with multiple sclerosis. I had periodically worked with her over the years, but when she came to my office after a few years of not making appointments, I instantly knew something was wrong. She had always been a beautiful, physically fit woman, but now, she had lost too much weight and did not seem herself when I saw her. She confessed to me that she had recently been diagnosed with MS (multiple sclerosis) and had been gradually losing her muscles and her strength. The body can only heal if the right nutrients are present. I treated her with massage and nutrition, and her MS symptoms have resolved completely.

Would you say working with spirits has assisted you in helping to heal some individuals?

Absolutely. Sometimes I get intuitions that guide my treatments. Other times, the messages come directly from the spirits. I can talk to them directly, just like we're talking now.

An Urban Shaman Heals through Ritual

I never felt at home in my family or in suburban Cleveland, where I was born and grew up. I felt I was different. Before I was old enough to understand myself, I could sense it. When I was eight or nine, my teacher gave our class an assignment to go home and write a poem. In my gut, I knew that I would not *write* a poem; I would *receive* a poem.

That night I felt the moon calling out to me. It wasn't with a literal voice, and I didn't see a spirit or a person; it was just a deep feeling I had inside, an intuition. I knew that if I went outside, the poem would just come to me. So I went out in the middle of the night, around 3 a.m. I wasn't afraid my mother would catch me because I figured the moon was bigger than my mom, and I would be okay. I went into the backyard and just lay down on the ground and stared up at the moon. Suddenly, and without explanation, words started to come to me. I began chanting and repeating what I heard over and over again in my head. When I was sure I remembered the poem, I went inside and stayed awake, repeating it until it got light enough outside to write it down.

That experience was a pivotal moment in my life. It was my first act of personal agency, really the start of what I can best describe as a guided intuition that has been present with me throughout my life.

I had a series of medical problems when I was a child. I was born with congenital bunions, a condition that is supposedly not possible. One of my parents' friends was a podiatrist and designed a painful device to correct my feet, but the device was excruciating, and I lived in daily pain. My mother dragged me to many doctors trying to find a permanent solution for me. None of the doctors thought surgery was a good idea, but my mom kept on searching, and she eventually found a doctor willing to try to fix me with surgery.

I had a series of surgeries; the first one was on both of my feet at once. I also reacted to the ether and needed to have my stomach pumped. My mother was sitting next to me when I came to, and she told me, "This hurts me more than it hurts you." I didn't know any curse words at the time, but what I was feeling in my mind was *Fuck you*.

From that experience, I realized that you're born alone, and you die alone in this life. That's all there is. I have me and my spirit's purpose, and that's it. The doctor who performed that first surgery insisted on doing multiple repair surgeries that ultimately left me with split toes. In many cultures, when a shaman is preparing to die, they look for an apprentice. They often look for a little child with split toes, which is a sign of spirit in those cultures. To this day, I have split toes. I came to realize that those surgeries were the beginning of my training as a shaman because in nearly every culture, a shaman's training includes isolation, pain, fear, and drugs, all of which I experienced.

When I was about thirteen, I woke up early and told my brother that I had dreamed that UN Secretary-General Dag Hammarskjöld had died in a plane crash. Later that day, the TV news announced a deadly accident had indeed happened. I can't speak about other people and how they have their gifts, but I know I was born with whatever I have.

In my early thirties, while I was in Los Angeles, California, and working on a book about a healing residency I had done at Manhattan

Psychiatric Hospital, a friend asked if I had a spirit guide to help me in this spiritual work. I told her I had always felt guidance in my life that I couldn't explain, but I didn't know if I had an actual spirit guide. She told me that she would talk to her mother and ask her.

I had no idea what my friend actually meant. Her mom had been dead for twenty years, but she still felt her mother's presence with her and would talk to her through a Ouija board. She took me to her home, and we sat at a table where she set up the board and placed my hand next to hers on the planchette. She began asking her mother questions, and the planchette began flying all over the place. I wasn't moving it, I know that, and I didn't feel like she was pushing it, either. Eventually, she asked, "Does Elena have a spirit guide?"

The planchette stopped moving entirely for a moment, and then it began moving differently. It moved slower, at a more deliberate pace, like a different entity was now moving it. This time it spelled out the word Kanin. "Is Kanin my spirit guide?" I asked. The entity replied, "Yes, but you should do more automatic writing because sometimes you're hard to get a hold of."

Kanin "told" me that we've had three lives together. The first was two million years ago in Siberia. Before it was known to the rest of the world that humans lived there, Kanin was with me. I was curious about Siberia and looked in an atlas, and there in the very north near the Arctic Circle is a peninsula named Kanin.

My spirit guide indicated to me that it was there where we first met. We've also had a life together in Wales, where we were "sisters but not sisters," meaning we were best friends though unrelated by blood. We were also together in the 1800s in the American Southwest. Every time I've had a session with a shaman or curandero, they've recognized that Siberian life. I did a past life regression ritual with another shaman, and she described in detail everything about that life in Siberia. They tell me things I'd only seen in my mind but never fully understood, but these things began to materialize in my "real" life.

I was at the Taos Pueblo many years later and went into a drum shop. When I walked in, a Tewa man and I recognized each other immediately. It was a compelling exchange. Later, I drove further onto the reservation land and had to stop my car after I became overwhelmed by emotion and started sobbing uncontrollably. Something terrible had happened in that place. As I learned much later, the Bureau of Indian Affairs had banished the tribe from Blue Lake, a sacred ceremonial center for the Tewa people, many of whom were massacred trying to defend it. I had died there in the past.

A short time after my experience in California, I found references to the chants of a famous Mazatec shaman, Maria Sabina. Her chants were so like my own they made me cry. Her power seduced me, and I read everything about her I could find.

Years later, I was visiting a friend in Geneva who lived with a Mexican man named Mario. As we talked, he told me about a woman I should meet in Mexico, a shaman who was his mushroom guide. The more he spoke of her, the more I realized she would be someone who would be good to meet, but I never thought I would travel to meet her. I stayed in contact with Mario over the years, and about ten years after we first met, I decided to go to Mexico. Since our initial meeting in Geneva, he had returned to Mexico and become a diplomat in the government. He told me he wouldn't be able to take me to see the shaman on this visit, but that I should see her, anyway. He said he would give me a letter of introduction to the lady who ran the café in the village closest to the mountain where the shaman lived. He assured me that someone there would help me to see Maria Sabina. Oh. My. Goddess! What a miracle! I had been connected to her through spirit all these years, and now I would finally meet her in person!

It took eight "pig and chicken buses" to get to the village nearest the rainforest. I knew this would be my test. And, sure enough, the last bus driver woke me from a nap and tried to rape me at knifepoint. But he was so drunk that when his pants were down around his ankles, I just pushed him down. All of his cronies had been standing outside looking through the window, waiting for the show. They became hysterical with laughter and taunts after I pushed him off me. He was embarrassed, and that was that. I got up and went inside the café.

The woman was expecting me. She said she'd gotten the letter from Mario and that the man who would take me to the shaman would be back in a couple of days. "He will take you up the mountain in his four-wheel-drive jeep and translate from Mazatec to Spanish for you." I waited in a room near the café, and on the third day, the man returned, and we went on our way.

When we arrived at Maria Sabina's hut, her daughter showed me in. Maria Sabina was the tiniest, oldest person I'd ever seen. She was lying in her bed, which was a pile of rags on the earthen floor. She held out her arms and embraced me. I recognized her immediately, and she recognized me as well. "I've been expecting you," she told me. She gave me the job of turning away the people who kept coming by foot and truck to be healed by her. My Spanish was good enough to deliver the message that she no longer had enough energy to help them.

We talked through the translator, and mostly, I just listened. She said she was dying, and that made her sad. I told her she would live forever in the hearts and bodies of all those she had healed and that I would carry her spirit with me always. She hugged me tight to her breast and whispered her magic into my heart. The feeling of exchanged energies was palpable. She was leaving her earthly power as I was coming into mine. After she finished this ritual, I felt a surge of energy pass through me and indeed felt as if she had transferred some of her extraordinary gift to me.

Then she blessed me to go home and do my shamanic work. Not to do hers—she had daughters, granddaughters, nieces, and apprentices who were Mazatec and would carry on her practice. I am not Mazatec and would never put on someone else's culture like a garment. I am what I am. I'm a modern city woman, an *urban shaman*. Shortly after our time together, I left the village and began my journey back home with a renewed sense of purpose, clarity, and awareness that ensured my path as a shaman.

When I was thirty, I had an intense vision of what I call *my assignment from the universe*. It connected me on a very visceral level to the reality of quantum physics. I instantly understood viscerally that everything is connected in an unfathomable web of energy, connection, and interconnection. I understood on a very intuitive level that creating a sense of community between people and connecting people to their own best selves, with the entire cosmos through ceremony, is among the noblest work that can be done. That is why it is so important to me.

I now do both private and public rituals to assist those who come to see me for aid. One of the things I'm most proud of is being what the *New Yorker* magazine calls me, "The unofficial Commissioner of Public Spirit of New York City."

For forty-five years and counting, I have led hundreds of free, public, participatory multicultural rituals for the equinoxes and solstices. These amazing events draw massive crowds of earnest participants in New York City.

Author: Do you have an altar in your home?

Yes, I do. I have many, at least one on every horizontal surface and every wall. Also, there's a connection with La Virgen de Guadalupe. I am a devotee of the Feminine Divine. I invoke and inspire that loving, nurturing energy in both women and men. I promote reverence to the divine feminine to help heal our planet from testosterone poisoning.

What types of consultations do you do with people? How did you get started?

I got started with tarot cards. Now I do not use traditional cards. I use a contemporary deck, and I never make predictions. I believe in free will. I use rituals and ceremonies to help people with all sorts of problems. Usually, I work through a three-step process of (1) owning, (2) cleansing, and (3) re-empowering. Ritual and ceremony are used for people to own or accept their current problems or challenges. Cleansing helps remove toxic energy and thought patterns that lead to difficulty. Re-empowering restores proper relationships and thoughts that will lead to greater connection and spiritual awareness. I can often feel what other people feel and know what problems they are dealing with before they even start speaking. I come from a "tough love" school of shamanism. I push people to grow and expand, and to be open to experience. I don't allow a client to persist in self-pity.

I also do rituals for protection. Sometimes people come to me looking to protect themselves from magic and other things they feel are dangerous. Often, they have difficult problems and are not sure what to do. When people come to me, they are often overwhelmed, frightened, driven crazy for all kinds of reasons. Sometimes people come just to get inspired, and I can help with that, too. I don't see spirits or speak with them, but I know they are real. I work primarily through my intuition and the cards. I use techniques from various cultures and ordinary things like herbs and branches that can be used for uncommon purposes. I do baby blessings and animal blessings and house cleansings, among other things. I don't do dark things, like hurting people or helping people get someone to love them. I don't put spells on people. Why would you even want someone who doesn't want you? I don't do that kind of thing. When someone feels unsafe, like they're being targeted, I can help them protect themselves through amulets, rituals, and other things.

Bembé and the Power
of Unseen Entities

When I was a child, I had many questions about the things I would see my family doing. I saw *bembés* at my home and my grandparents' homes. I can understand how it would be easy for someone unfamiliar with Santería to have questions about *bembés*— the things that happen during them are intense. People would come from all over Cuba for our *bembés*. The house would be full of neighbors, family members, friends, and other people my family knew, and after everyone had arrived, we would head to the backyard. The guests would always have a bottle of rum, sweets, a live chicken, or some other gift with them. You never go empty-handed to a *bembé*.

I would watch them as the *bembé* began, and it would always start with the touch of the conga drums. My father is a son of San Lazaro, and the *bembés* I attended would always be done in honor of San Lazaro. The drumbeat for San Lazaro is unique. Only a few people know it, and it would commonly be the same drummers who played the conga at our *bembés*. The priest would usually either pray or sing praises, and the people would start to dance for San Lazaro. Some would be dressed in sackcloth, and some would have a red handkerchief or a headband made from torn red cloth around their head. As the music continued, many people would go into transition, sometimes even the priest.

While the santero was in transition, San Lazaro would express his presence through the various attendees' works. Different individuals often manifested San Lazaro's presence in different ways. My great-grandfather, for example, was always in transition at the *bembés*. He wasn't a shaman or a curandero, but spirits would always bring him into transition during rituals. I would watch as his voice, movements, and mannerisms would change so intensely that he seemed like a completely different person.

The priest would manifest San Lazaro through healing, clairvoyance, and divination. He would walk through the crowd and ask people to place their hands over where they needed healing. He would stop at each person and address their concerns, often giving details on specific problems the individual had without being told about them. Those in need of healing would be treated with a touch or a prayer. Some came with serious problems.

It's difficult for me to believe some of the things I've seen at *bembés*, so I can easily understand why people have so many doubts and questions about the saints. I have seen people drink entire bottles of rum in seconds and not become drunk, seen people eat pieces of glass without being cut or injured, and seen elderly women picking up men twice their size. I've seen people put out cigars and other burning things on their skin and not get burned. It's also common to see people hit by a machete and not be injured. I've seen other things that stretch the imaginations of those unfamiliar with the powers of the spirits. You have to see to believe.

The altars are for San Lazaro. The saints are fed traditional foods like coconut nougat and rum, and cigars are placed on the altar for the saint through rituals. During the *bembé*, the blood of a goat or chicken is often used as an offering. In return, the saint will fulfill promises and grant wisdom, knowledge, and healing.

When I was young, my mother told me about how many Africans came as slaves to Cuba. With them, they brought their religion and their beliefs. The Spaniards did not allow the African slaves to practice their faith, but many of these slaves escaped to the mountains as they grew to loathe their masters and slavery. Eventually, these runaway slaves formed quilombos, communities of fugitive slaves with houses hidden in the mountains. As more slaves escaped, the quilombos grew. The escapees used medicinal plants and animal products and their religion to heal the sick in their communities. The fugitive slaves also became engaged in honey bee harvesting and traded honey to poor Cuban farmers for food. When some of the farmers or their children became sick, they often went to the quilombos for healing. Doctors were unavailable or too expensive for the peasants to see.

Within those quilombos, the holidays of the African Yoruba religion were celebrated. Over time, the peasants and quilombos intermarried, and the African religions and their beliefs were transmitted to the Indigenous Cuban people. Santería took form as a blend of Catholicism, the Yoruba faith, and Indigenous Cuban beliefs as the slaves and peasants attempted to hide their rituals from the predominantly Catholic Spaniards, who vigorously tried to prevent slaves from practicing their religion, often in vicious ways.

During *bembés*, after the saints' healing and wisdom were shared, the men and the priests would exit the outdoor area. The drumming continued, and the men playing the drums would be shielded from the women by a curtain or a wall. The devoted women then danced naked for San Lazaro. It is a lovely and pure ritual!

Older members of my family say that the *bembés* are no longer the same. Sometimes people use the occasion just to become drunk. The most dangerous thing is when you make a promise at a *bembé*, you have to keep it. Only make a promise you can keep. Sometimes people don't keep their promises, and the saint makes them sick, drives them crazy, or rots their skin or something. If you do not like this world, do not get into it. But you should respect it.

I grew up with *bembé* and saw lots of different rituals. Some were done for healing, some for protection, some for fertility, some for other reasons. As I grew up and studied various religions, I found a spirituality that works for me. I don't go to church or temple; I pray and read holy books.

Because of my past, Santería will always be a part of me, but there's a part of me that's different. It's how I understand God. For me, with God, there's nothing that he asks back of you but to be good. There are no rituals, no giving gifts, cigars, rum, blood, or anything. All you have to do is pray from your heart for good things and be a good person. I never ask for things for myself. I pray for my family, my children, and ask God to help me be a better person. Santería will always be a part of me, but there is something about the spirits asking for things in exchange for performing acts that doesn't work for me. I know God wouldn't act that way.

The Discovery of Alien Life
Reflections on Unseen Entities
and *The World Hidden*

*"The intuitive mind is a sacred gift and the rational mind is
a faithful servant. We have created a society that honors the
servant and has forgotten the gift.*
—*Albert Einstein*

*"Intuition is a very powerful thing,
more powerful than intellect.*
—*Steve Jobs*

What the hell happened to CNN? The network used to be way more toward the center when Ted Turner was the owner. Fox has always been hard to the right, that's to be expected from them, but CNN has become the anti-Fox, a twenty-four-hour anti-Trump network now, it seems, with little more mentioned than anti-Republican politics and the coronavirus. Hello, McFly?! Hello?! Other important things are happening around the globe every day, and yes, they deserve some quality news coverage too. Don't get me wrong, I still enjoy the network, I do. Fareed Zakaria is a very fair journalist and has one of the best shows on television. The CNN anchors are still some of the best on TV to me, and Chris Cillizza is one of the best-writing journalists I've ever read. I usually watch an amalgam of CNN, Fox, and Al Jazeera a little every day, hoping to piece together the puzzle of what the truth of what's happening in the world really is.

Journalism is a beautiful and incredibly powerful medium. In recent years, particularly in the age of Trumpism, the media has come under significant attack from both the political Left and Right—and not without merit. It's difficult in this era of corporate addiction to TV ratings and extreme political polarization to find pure and unbiased journalism. Information gathered honestly and with integrity and stories focused on global societal impact, told artfully without some hidden agenda, are indeed rare gems. Such is the modern journalistic environment, and it can be easy to dismiss this beautiful medium, an under-appreciated art form in my opinion. Still, there is incredible elegance and power in telling the truth in a way that moves people with the essence of honesty. And if you don't believe journalism can be art, take a look at a few episodes of Anthony Bourdain's *Parts Unknown* and we'll see if you still feel the same way.

How do you convince people of a truth they cannot see? That was the central question I asked myself after I became certain unseen entities are real. It was daunting. Still, I felt, and accurately so, there must be some quantifiable way to demonstrate what people were subjectively describing to me about the existence of these unseen entities.

The world witnessed the power of journalism as a medium when we watched the death of George Floyd play out on televisions, phones, tablets, and computer screens around the world. The documentary footage of Floyd's cruel and callous murder sparked protests throughout the US and other countries. That is the true power of journalism.

The video of Floyd's murder couldn't be spun to paint a different picture. There was no hidden guise within the video and no way to convince the world that what it witnessed was anything other than what it was: a heartless murder orchestrated by a member of the police. That is the power of journalism at its heart. Documentary video, as photojournalism did before the advent of video, cell phones, and social media, elevated journalism to a uniquely more powerful plateau. If the famous often-quoted mantra "a picture is worth a thousand words" is true, then how many words is documentary footage worth?

When I wrote my first work, I knew the book and its accounts would come under intense scrutiny. I was sharing accounts that appropriately should be met with great skepticism, as many of the accounts stretch credulity. I was postulating the existence of intelligent alien life forms that cannot be seen except by a few among humanity, something that even I would have difficulty believing had I not personally experienced the power of genuine shamans. That is a declaration of unquantifiable significance and consequence—not just for science but for religion and humanity itself. Such a proclamation should be met with the highest degree of scientific scrutiny, and every effort should be made to verify the accuracy and veracity of such accounts.

The World Hidden was born principally because I lacked the means to do what I really wanted to do when I first understood the significance of unseen entities and their activity with humankind. Initially, I wanted to tell the story of Indigenous healers through a blend of documentary filmmaking and scientific study and by monitoring the biometrics of these healers' brains and bodies. Brain analysis while in transition (a term used to describe an unseen entity's direct interaction with someone), radiographic imaging (PET, MRI, and CT scans), the analysis of patients' blood chemistry before and after treatment, DNA analysis of generations of healers from various locales globally, and complex and sophisticated analysis of the water from these healers' altars form the principal foundation of the research work I want to do. After forming a friendship with a shaman and visiting with her over an extended period, I noticed that the water in the wine glass she used to perform consultations changed. It took on a unique opaqueness, an apparent alteration in its chemical composition that does not happen when water simply sits in a wine glass. The morphology of the water in the glass also appeared unique, with a perfect sphere just below the surface of the water developing over time.

Water is a commonly used tool among many shamans as a means for interacting with unseen entities during periods of clairvoyance. It is thought to sequester and hold these entities. This is a nearly uniform custom for many shamans and others who interact with unseen entities. In time, I grew convinced that a simple but thorough analysis of this water would conclusively, scientifically reveal the truth of these unseen entities; it is also relatively easy research to do if one has access to cooperative genuine shamans and the financial means to do so. So, I began my work in this area and in a short time saw the accuracy of my initial assumption. *Shamanic water is the principal way to demonstrate the reality of these entities' existence.*

Researching Unseen Entities

Unseen entities interact with the physical world in three principal ways, and it is by examining these three ways the evidence of their existence can be established beyond all doubt.

I. Analysis of Shamanic Water

Unseen entities are sequestered in the sacred space of shamans and others with the ability to interact with them through the use of water. This water is then used to perform consultations with clients seeking assistance. Nothing can effect a change in the physical realm without leaving some trace of its existence. Unseen entities are no different in this regard.

The water of shamanic altars is altered morphologically, principally because these beings are composed of a type of energy that modern scientific study has not yet been able to fully identify, assess, and understand. The development of a perfect sphere in this water is the most obvious change that can be readily observed and studied, and comprises nearly entirely the foundation of evidence I've garnered in my research and observations. Studying the *why* and *how* of this change is complicated and demands a multidisciplinary approach. Variations in the conduction of electricity; variations in light and laser transmittance; and the effects of temperature fluctuations on these and other factors, including spectral imaging, electron microscopic imaging, and other expensive and complicated physical assessment tools useful in scientific evaluation, such as gas chromatography and mass spectroscopy of the water before and after shamanic intervention, form the foundation of the multitude of research experiments that give a more complete picture of how this water is altered by the presence of these entities.

II. Biometric Analysis of Shamans in Transition

The physical and chemical examination of water, though essential, is not enough alone to establish that unseen entities are responsible for the change in shamanic water. It is vital that three levels of experimentation occur simultaneously to establish beyond all doubt that the change in water, the change in shamans, and the change in individuals in consultation with shamans are all related and interconnected.

Transition, a term often employed to describe the period when shamans channel unseen entities into their bodies, is characterized by changes to both their physical and mental states that can be monitored, measured, and evaluated. When shamans enter transition, they lose conscious awareness and cannot remember anything that these entities say and do during this period. The period before conscious awareness is lost, but when the process of the entity gaining control of the shaman's faculties has begun, is characterized by physical changes in the shaman. Frequently, changes occur in the motor control of the tongue, and the shaman's speech is affected. Such changes in mentation and physical muscular changes likely will have corresponding EEG (electroencephalography) changes. There are also likely neurohormonal changes and other biometric changes that can be observed with spectral imaging, likely linked to the changes observable in the water used in shamanic rituals.

Another interesting pattern described by the large collection of individuals I've interviewed is what they describe as a period of awakening after the entity is allowed to interact fully with them and gain control over their faculties. The immediate period after transition is characterized by a change in the shaman's visual perception, where colors and other physical sensations are enhanced. Colors appear more vivid, for example. The long-term effect of transition is that shamans are from the period after the initial transition able to see the entities the same way they are able to see other physical objects. This change suggests that transition alters these individuals' brains in some permanent way that perhaps alters the visual spectrum their brain is able to perceive. This is an alteration that must be assessed and studied to be fully understood and comprehended.

III. Biometric Analysis of Patients

The third and final crucial component of analysis that must be incorporated and coordinated with the simultaneous analysis of both shamanic water and shamans is the biometric analysis of patients interacting with shamans. It has become obvious to me that unseen entities are composed of, and operate principally through, the use of a type of energy that humankind has little understanding of yet. That energy is transferred through shamans during the work they accomplish, and it will have some measurable consequence in the physical world that I believe is observable in shamanic water, shamans, and those toward whom such energy is directed.

Many types of healing feats are reported by the shamans and others I interviewed in *The World Hidden*. Biometric analysis, spectral imaging, and radiographic study prior to, during, and after shamanic treatment intervention comprise the principal ways to monitor patients treated for conditions with objective measures that can be readily monitored. Autoimmune markers; tumors, particularly inoperable ones; hypertension; and other conditions offering concrete observable metrics can show whether any therapeutic intervention has any consequence on such conditions. Other psychiatric conditions such as PTSD, drug and alcohol addiction, schizophrenia, and depression are all predominantly subjectively assessed in the sense that what patients describe they experience largely provides guidance about the quality and effectiveness of any given treatment, and these should also be included in evaluation. Various assessment tools are commonly used in research studies to measure the effectiveness of various treatment modalities. Psychological conditions should also be included in research studies, for perhaps in no other field of medical endeavor is the need for innovative and effective treatments so great. I suspect shamans and unseen entities work primarily through the manipulation of the energy of intention, the same energy that allows us to sense when others are staring at us though our back is turned, for example.

There were many things I saw with my own eyes and others that were relayed to me that I intentionally left out of *The World Hidden*. *Words are just not enough sometimes.* Many of the accounts I shared I knew would be hard to believe, and I didn't want to challenge the credulity of readers beyond a certain point. Many of the book's contributors, including my wife, were exceptionally candid and forthright in sharing their experiences with me. I knew many of the experiences I excluded from the book would simply be met with ridicule, even beyond how the book is already scrutinized. My wife recalls, when she was a young girl, watching her grandmother make it rain and stop thunderstorms by just putting an iron hammer on the ground. I've personally observed the manifestation of shamanic shapeshifting, and there is simply no number of words, no matter how artfully crafted, which will convince people of the reality of such things without showing them. Documentary filmmaking combined with organized scientific research is the only reasonable path forward to demonstrate what shamans and others are able to do through their interaction with unseen entities.

A few years before my transformative experience, and not long after I moved to New Mexico, I pursued a couple of unique entrepreneurial endeavors. I'd written the rough outline for a sci-fi story that featured an Indigenous couple as heroines. I loved the story outline but let it linger for a while before writing it. At the time, I was working in an office practice and in urgent care, and I would play more video games than I should on the weekends and at night after my long shifts, during which I'd sometimes see sixty patients or more. I'd recently met a patient who was a genius computer programmer, and he'd developed some of the most famous video games of all time and other simulation programs for the military. We talked and got to know each other. He loved the idea of the story; I presented him with some novel ideas for gameplay and replayability that would enhance the game, and we got to work. I further developed the story and had storyboards made. We recruited local artists to help make 3D models of characters and environments. The artists were volunteers, students from the local university and community college, and though they were talented, it's often difficult to get the required quality and rate of production from students without professional experience. Eventually, the project stalled.

A few months later, I began thinking about other projects the programmer and I could work on together; this time, I chose something educational. I had an idea for interactive textbooks and medical training software that would blend augmented and virtual reality to make an innovative teaching tool for medical students. The idea would be to train students to think like doctors and to present training scenarios for complicated medical procedures that are otherwise only learned on actual patients. The programs would expose medical students to common emergencies, for example, how to deliver a baby, put in a central line, intubate, do nerve blocks, and other common procedures and scenarios most doctors should be comfortable handling. Heart attacks, aneurysms, strokes, gunshot wounds, vehicular trauma, and other frequent emergency and intraoperative complications would be part of the training. The students would handle these scenarios repeatedly in virtual reality and train for procedures with a combination of virtual and augmented reality with cadavers and dummies so that when the time came to complete the procedures on actual patients, they would have considerably more familiarity with the procedures and scenarios for which students and residents customarily get little to no training prior to real-life situations.

The point of the innovative program was to reduce the likelihood of error, which, in my opinion, is a significant problem with medical training in that much of what must be learned is typically predominantly done on actual patients. This presents obvious problems, and anything that can reduce patient risk and enhance student learning is worth at least considering. I presented my idea to a local medical school, and it was universally accepted as practical and useful, but because of the significant cost associated with developing the program, it was dead on arrival. I'm not the sort who has a cadre of wealthy venture capitalist friends on hand to front me the capital for a product that would require millions of dollars of upfront investment to develop, so though the idea was good and practical, it also stalled. I placed it on my list of failed inventions and other ideas that went nowhere because of a lack of capital.

Looking back, the disappointment I felt from the experience of the failed video game and medical training program helped me realize the importance of not aiming too high initially when trying to achieve something of great significance. I refused to make the same mistake approaching unseen entities as I had made with these and other projects. I deftly know my way around a camera, but I'm no cinematographer. The documentary I want to do is international in scope and sophisticated in nature, encompassing medical imaging, biometric analysis, and basic science research. I completely outlined the process and thought deeply about how to proceed; this time, however, instead of shooting for the moon, I chose to start with just writing a book, a far more realistic goal. I was driven and had the help of many benevolent shamans and others they had helped. I also realized that I needed to let the people speak, which I believe is where the true power of *The World Hidden* lies.

The World Hidden's appeal, in my opinion, is that it is principally a work of journalism. There's some modest commentary by me, outlining the societal ramifications of the work shamans and others who interact with unseen entities do, but the book's true beauty is in the accounts of the people who shared their stories with me. To my wife's dismay, I simply let the people speak and didn't try to craft conversation and narratives too elegantly. There's tremendous power in preserving the authenticity of a person's experience. My time as a photographer and photojournalist taught me that, and a lot about dealing with people in general. A camera is an intrusive instrument, and honest documentary photography requires accessing people's vulnerability.

When I was a senior at Morehouse, not long after my first exhibit debuted, a respected friend asked me if I thought my work was exploiting the impoverished people I was photographing. I looked at him, thought about it a moment, and answered truthfully that it in some sense, yes, it does. I then sincerely responded that the intention behind the work was noble. I was convinced of the importance of showing a side of life that is often left unseen or shown in a way that lacks nuance, showing only the ugly side of poverty. I also gave the people I photographed a significant portion of the money I earned from the exhibit. There's never been a harsher critic of my work, both ethically and artistically, than I, and my conscience was clean knowing that my intention and my manner of practice showed both respect and gratitude for the individuals who let me capture a part of their lives.

I've always, and without exception, gotten permission before I photographed anyone and still do to this day. I followed the same principles in collecting the accounts shared in my written work. I've never shared anything that any person granting me access to their life didn't approve of me sharing, and I never will. The shamans and others who interact with unseen entities, I'm sure, knew that I would respect them in this way and respect any boundaries they put on what I could and could not share. This in no small part contributed to their candor. As a journalist, particularly a photojournalist, the aura you manifest and the spirit with which you engage in this type of intimate work can be felt by those whom you photograph and other observers, even those without the unique hyper-developed intuition and clairvoyance of shamans. That aura, one that should display a commitment to fairness and integrity, humility and confidence, lies at the heart of encouraging individuals to be open and honest when sharing things on the record.

For centuries, documented accounts of Indigenous people have presented various abilities shamans manifest as a result of their ability to interact with unseen entities. Aside from the analysis of shamanic water, I didn't present anything the world hasn't heard before. For me, the beauty of my work rests in the words of the people who shared their accounts and in my relating, through diligent verification, the veracity of those accounts. Masterpieces like *El Monte*, Eliade's *Shamanism*, *The Book of St. Cyprian*, *The Keys of Solomon*, and other ethnographic and occult masterpieces have outlined in far more detailed and elegant ways the depth of ability shamans and others who interact with unseen entities manifest. The variety and diversity of thought regarding the ability of unseen entities is presented far more elegantly in these works, and those who through various means interact with and use unseen entities to perform a variety of difficult-to-believe feats are explained in significantly greater detail.

When I first shared some of my writing with my wife, I asked her what she thought of it. Her cryptic response was, "You write like a man"—not exactly the feedback I was hoping for. I wasn't exactly sure what she meant at first, but the tone of her voice and the look on her face weren't exactly encouraging. I compelled her to explain what she meant. "You're logical and you express yourself clearly, but there's no color, no flamboyance in the way you wrote it. It feels like a journalist telling the news." I couldn't deny it. She was right, and I laughed as I objectively considered her words, but that was exactly my intention. I was trained as a journalist and never really broke that training. My first editor shared my wife's opinion of my writing and had no qualms in repeatedly telling me my writing was shit. She never explained why or exactly what she didn't like about it; she just wasn't a fan. Some writers are master storytellers; others, exceptional wordsmiths; and the truly rare ones like my wife and Jose Martí are a combination of eloquence and profundity, whose range is broad and includes everything from political essays to poetry for children. But *The World Hidden* isn't about me or my writing, it's about the people who honestly and candidly shared their accounts with me, and I stuck with a simple and straightforward tone.

My wife is an incredible writer with a foul mouth and a spectacular sense of humor. She just has a way with words and with people. Her style and way of connecting with people simply cannot be taught. She naturally has something I'll never achieve, no matter how long I work on my craft. But she writes to entertain, and I was writing eyewitness accounts, trying to match the tone of how the individuals told them to me. We're opposites, my wife and I, both as writers and as individuals. When she read my writing in *The World Hidden*, it wasn't her choice of style; still, she was convinced my work would sell and that I'd become a successful writer in time because of the depth of my ideas and the significance of my observations. In a way, her reaction often illustrates the ability to distinguish something that's entertaining from something that is practical and useful.

The World Hidden is not a book many people would usually even consider. The man who wrote the foreword for the book is a close family friend. As he is a physician and an intellectual with a lifetime of distinguished accomplishments and medical credentials, I asked him to write the foreword because I knew he would not only do a good job but also be completely honest. As a conservative Christian and an exemplary physician, he represents both segments of the populace I felt would be most resistant to the book's ideas. If he found the concepts and conclusions insignificant, it would be worth hearing his objections, and if he saw the merits of the accounts and theories, I had hope others whose inclination would be to dismiss my findings would also find value in the book.

His response, along with those of other friends and family members, convinced me I had indeed found the middle ground: a book that had significance and one that could be understood and considered by people from all walks of life. I revisit some of those same concepts here, with the hope that the obvious merit of my work and its profound ramifications for humanity will touch the minds and hearts of the readers.

Are Unseen Entities Actually Aliens?

Are unseen entities, commonly referred to as spirits, actually a form of alien life? Why have I chosen to refer to them as alien life forms? For many of the healers I've spoken with, unseen entities are far more than aliens—for some, they are deities, and the rituals done to gain various entities' favor are acts of worship. For others with the ability to interact with unseen entities, these entities are not deities but the remnants of dead loved ones, something left over after we die. For still others, they are simply intelligent nonhuman beings, some of which are benevolent and others malevolent. For some, the rituals done to interact and gain favor with unseen entities are simply an exchange in which the entity is granted something desired in exchange for some favor or wisdom the person interacting with the entity is seeking to gain. Rituals to gain protection, healing, clairvoyance, and guidance for a particular problem are routinely done by shamans and others who interact with unseen entities.

I have chosen to describe unseen entities as forms of alien life for several reasons.

1. There is a great divergence of opinion among shamans and others who interact with unseen entities as to what these beings actually are. A nonspecific term is helpful when attempting to classify these entities, since it is still unclear to me what they are scientifically.
2. Unseen entities show obvious signs of superior intellect and powerful ability and are nonhuman. Those with the ability to see and speak with them report having conversations with these beings that are often diverse and varied.
3. For too long, the existence of unseen entities has been entirely shrouded in the realm of religious mysticism. Religion, television, and cinema have nearly entirely defined how society views unseen entities, often showing only the destructive and malevolent aspects of some entities, completely ignoring the accounts of shamans and others who interact with unseen entities about their ability to cooperate with these entities for the benefit of others.
4. Referring to what many cultures commonly name "spirits" as unseen alien life forms helps bring an objectivity that has been sorely missing from society's consideration of unseen entities, an objectivity that is essential for sincere scientific evaluation, study, and classification and comprehension of the risks, dangers, and potential benefits that these beings pose.

Because there's never been mainstream acceptance of the reality of unseen entities, there's never been the type of organized and detailed study and research these beings merit. While some work has been done in the shadows and in secret by a select few, without the widespread acceptance and acknowledgment of the existence of unseen entities, the type of mainstream study and research that is needed to fully understand them, what they are, what they are capable of, where they come from, what risks they pose, and so forth is not fully understood. For centuries, alien life forms have existed among humankind, yet very little is known and understood about their actual nature and abilities. It is time for modern society to accept the reality of unseen entities and remove the shroud of ignorance that's prevented the world from having insight into a true form of alien life that surrounds us daily, though unseen by the majority of humankind.

The Gift

Many of the individuals I've come to know who interact with unseen entities speak of their ability to see and communicate with them as having *the gift*. For some, the ability to communicate with unseen entities manifested in childhood, when the children were around individuals in their family who also had the gift and/or were participating in rituals with unseen entities. From the information I've gathered, I've learned that for a separate, small minority, their ability to interact with unseen entities occurred with no outside influence; that is, no one in their family or around them had any relationship with these entities or any ability to interact with them. For others, particularly for some participants in Santería, they had no natural ability to interact with spirits, but only after paying a santero—in some cases, up to US$25,000 or more, and then only through a series of initiation rituals—did they gain the ability to communicate with and see unseen entities.

One of the curanderos I spoke with believes these entities will interact with an individual only after that person has shown an interest in them. Whereas another curandera born with the gift stated that it is the entity who chooses the individual, often with the individual showing no interest and refusing to interact with the entity for years. This curandera also believes that the gift is something that runs in a person's blood and sometimes skips a generation.

Who is right? Perhaps they both are. Perhaps for some, the gift is something they are born with and an entity chooses them without any interest being shown on their part, whereas for others, once they gain knowledge, experience, and awareness of the reality of unseen entities, their personal interest brings one or more entities to them.

DNA analysis of healers born with the gift and families in which the gift has manifested through several generations is a significant research interest for me. Still another possibility is to simply interview an unseen entity. Having a focused interview with an unseen entity to gain their input on a host of topics is a subject that fascinates me. Certain reference materials on unseen entities speak of ways in which a person can discuss any topic with an entity and gain input from them. For centuries, such knowledge has reportedly been hidden and used primarily for the personal gain of the individual possessing such knowledge. Perhaps in the future, the wisdom and insight of unseen entities will be gathered on a host of modern subjects, particularly areas of universal societal importance such as health, the environment, religious faith, and interpersonal relationships.

The Ethics of the Gift

One of the most fascinating aspects of the world of shamans, paleros, and others who communicate with unseen entities is the unique code of ethics that often guides the work they do or, in many cases, refuse to do. There is enormous variation in the manner and nature of the work various healers and others do with unseen entities.

Many shamans, paleros, curanderos, santeros, and others who work with entities exclusively for healing often hold themselves to very high moral and ethical principles. Frequently, there is a period of celibacy for curanderos and santeros, typically a year, during which the individual focuses on cultivating their spirituality and denies themselves sexual intimacy to enhance their ability to interact with unseen entities.

There is also a common, nearly universal principle these healers have of honesty. The overwhelming majority of healers I've met do not charge for the work they do. They often view their ability to interact with unseen entities as a gift from God and, as such, freely dispense their assistance, even though they may live in poverty and survive off the generosity of their community and the individuals they serve.

For those who interact with unseen entities for monetary profit, there is a considerable difference in their work and the manner in which they do it. For many healers, the thought of using their ability to injure others intentionally is repulsive. While the protection rituals healers do for others may result in injury to individuals attempting to harm a person they've protected, they see such injury as the result of the person's malevolent intent. Work is never done to intentionally harm except in unique instances.

For others who interact with unseen entities for profit, using their ability to harm or heal is often considered a transaction. The entity is contacted and asked what is needed to accomplish the task. The work is then completed according to the entity's directions, often with an offering or gift given to the entity in exchange for the task to be accomplished. Sometimes, entities considered benevolent are also used for this work, even work that has malevolent intentions. Such activity encourages one to look at unseen entities without absolutism. Sometimes, for example, malevolent entities can be used to heal and benevolent entities to harm.

Many interactions with unseen entities involve what I would consider a transaction, that is, the entity is granted something, be it blood, animals, rum, or some other organic material, in exchange for the assistance it is being asked to provide. What is the actual benefit these entities receive from such exchanges? Does the blood somehow strengthen the entity? Spirits are considered entities without physical form, and the words of various healers suggest that their power is not without limitations. One curandero mentioned that the entities he interacts with often must be fed before they can be used for healing. How is it that blood and other organic materials feed them?

By far, one of the most fascinating experiences I've observed is the *change of life* ritual. Here, the life force of one person, usually someone elderly and infirm, is transferred to an entity in exchange for healing another person who is seriously ill. The ethics of this ritual are complex and fascinating. In some instances, the person's life force is given willingly, making the ritual a form of assisted suicide. For others, the ritual is done without consent and can affect more than one person. At times, several individuals' life forces are taken unwillingly. This is obviously malevolent action by both the entity and the practitioner cooperating with the entity. It raises not only ethical questions regarding the nature of unseen entities and certain practitioners but also legitimate questions about the degree of danger unseen entities pose.

Part of the impetus for scientific inquiry is to understand the nature of unseen entities more thoroughly because, as much as healing and clairvoyance can be of beneficial use, the destructive power of unseen entities is something that should also be respected.

Militarization of Unseen Entities

The healing and other beneficial works performed by unseen entities are what primarily captured my interest and sparked my inquiry into the world of shamans and others who interact with what they often describe as spirits. My early exposure was to individuals who work with unseen entities for the benefit of others. As I became more informed and gathered accounts and experiences from various healers and others, I realized that other powerfully gifted individuals interact with unseen entities for the injury of others.

These individuals' work has tremendous societal and military ramifications and applications. And in many ways, the militarization of TOTI has already taken place. Despite my personal feelings and misgivings about such activities' ethics, I have salient thoughts about both the potential dangers and the usefulness that unseen entities and certain individuals who work with malevolent intentions have. What they pose to society can be harmful but can also be used in ways some may find of value.

It lacks nuance to think all who interact with unseen entities for malevolent intentions are evil individuals. Some who carry out such work consider themselves agents of justice and harm only those who have injured others seriously. Some follow the same principle as many healers by not accepting payment for their work and by not harming others who have done nothing the practitioner considers significantly evil. For others, without such ethical limitations on their work, their potential contribution to society is also worthy of consideration.

It's the nature of nearly every technological advancement to have the potential for both benefit and harm to society. It's not infrequent that some discovery's application leads primarily to the damage of humankind and the environment. Whether or not such use is ethical will not prevent some people from using their ability for their own financial profit, regardless of the potential harm to others. We are wise to consider the potential military applications of unseen entities' powers, as it is a reality that those abilities are often applied with malevolent intentions. In this sense, the militarization of this technology has already occurred.

Clairvoyance and the ability to directly injure individuals have obvious military applications. To know an enemy's plans, their whereabouts, if someone is a spy, and what an enemy's interests might be are crucial potential applications of interacting with unseen entities. Powerful individuals who work with spirits for malicious purposes can injure and kill from any distance for the right price. A tarot card reader, that is, one who claims to be guided by an unseen entity, described having the ability to know any information they desired about the present through the cards' use, including the ability to find lost persons and objects. Imagine knowing where an enemy's weapons facility is located or what weapons an enemy is developing.

Imagine learning these things without exposing other individuals to danger or harm and without the use of satellite or other technology. This ability is priceless and is a potential source of revolutionary military application.

Clairvoyant soldiers, bodyguards, detectives, and assassins are not a thing of imagination. The development of military technology as it relates to unseen entities, and individuals with the ability to interact with them, has obvious and tremendous value. It is likely to develop rapidly as science makes progress in understanding some individuals' ability to interact with unseen entities.

Many other potential applications have value. Frequently, the individuals working to harm others do so because they would otherwise live in poverty. Finding an alternative way for them to use their abilities ethically would be practical and beneficial to society.

The Dangers of Unseen Entities

Are unseen entities dangerous? Absolutely. Unseen entities are not all the same in terms of their ability or morality. This is the clear consensus of the healers and others I've interviewed who interact with them. There is also tremendous nuance when it comes to understanding unseen entities. Many individuals who work with them speak of both good and evil spirits. They also give accounts of benevolent spirits being used for malevolent actions and malevolent spirits being used for healing and other benevolent activities. Understanding the nature of the ethics and morality of unseen entities is a complex and rich avenue for potential study.

Unseen entities are vastly superior to humans in intelligence, power, and ability and thus have the capacity for more benevolent and darker, more malevolent actions. It appears the activities of unseen entities relating to healers and others who interact with them are commonly transactional. The entity requests something in exchange for the action or guidance it gives to humans, especially for more intensive requests made, such as in cases of serious sickness. At times, however, there is no transaction for the wisdom or guidance given, particularly with respect to spontaneous moments of clairvoyance.

There are also many instances when unseen entities harass, injure, and, in some cases, possess individuals, with significantly harmful effects. Several healers I spoke with described examples of harassment by malevolent entities leading to mental illness and even suicide. Sometimes, the healers were themselves the victims of harassment and disease brought by the very entities they eventually partnered with to accomplish years of benevolent actions. The harassment and infirmities inflicted upon them were done with the intent of coercing the reluctant healer into a cooperative relationship with the entity or entities.

One curandero I spoke with mentioned a family that was nervous about an entity that liked to make noises in their house. The noises disturbed the parents and the children, and the father asked the curandero how to remove the entity from the home. To my surprise, the curandero told me that this type of entity is not harmful but merely playful. He told the father that the entity was actually protecting his family and that they should not remove it but should learn to live with it.

According to some of the healers I spoke with, entities are like humans. They have a sense of humor and take pleasure in various activities, some harmful and some harmless. Some entities are thought to enjoy torture, blood, and sexual behavior that many would consider deviant. Some serial killers have spoken about being harassed by entities, and after hearing the reports of some of the individuals I've interviewed, I believe there is merit in some of these killers' claims.

In the section titled "The Ethics of the Gift," I wrote about a ritual called the change of life, one of the potential dangers involved with unseen entities. In that ritual, one or more individuals' life force is given to an unseen entity in exchange for the entity healing another person who is gravely ill. Mattias, a curandero in Cuba, described an instance of a change of life ritual he did to save his mother's life. During this ritual, a spirit asked to take the lives of several patients at the hospital in exchange for saving the life of Mattias's mother (account contained in *The World Hidden*). It is unclear how many patients died or were injured during the course of this ritual, but the result was that his mother was healed, as were several other patients in the ICU. I think it's notable that the entity did more healing than was asked, but it is unclear the degree to which others were harmed. This ritual raises many questions about the potential dangers of unseen entities.

Of all the rituals I've seen and been told about, the change of life ritual is perhaps the one I would most like to see studied and gain more understanding about. The ethics of this ritual are, without question, troubling, but in the context of assisted suicide, there is an argument to be made about its place in ethical medical practice. Regardless of the ethics of or how we may personally feel about assisted suicide, the change of life is a ritual that takes place and reveals a lot about unseen entities' nature. Does the phenomenon of an entity absorbing another person's life force occur in circumstances outside this ritual? If not, why not? What are the limitations unseen entities have, both ethically and physically? Are there measurable changes in blood chemistry, brain activity, and other biometrics of the individual(s) being healed and those being harmed during the ritual? Does the ritual itself reveal any other severe, hidden dangers unseen entities pose? There are numerous scientific and ethical questions that this ritual provokes.

It is not insignificant that unseen entities have coexisted with humans for millennia. Shamanic traditions in cultures worldwide suggest that benevolent spirits, as some describe them, have been sources of useful guidance, support, and acts of healing for the peoples and communities these shamans serve. The cultures that include shamanic practice are cultures where the people coexist in nature with a lifestyle that results in little to no sustained damage to the community surrounding them. The entities themselves are fed and sustained with organic material, and the healing direction unseen entities give is often rooted in the animal and plant life available to the shamans in their locale. This symbiosis leads to mutual respect for the unseen entities and the natural world, which provides the nourishment the entities request. This respect for the natural world affects the community at large and is something that modern society has significantly lost touch with, to considerably deleterious effect.

Reconciling Shamans, Spirits, and Science

Could unseen entities, commonly referred to as spirits, actually just be poorly understood alien life forms? Are unseen entities real? Yes. But how can unseen entities be explained scientifically? The truth is that intelligent alien life forms do exist and have been in contact with humans for thousands of years—but in a form and manner that have been shrouded in the realm of the mysterious and the religious. Many people around the world are convinced that unseen entities are real, and there are very legitimate reasons why.

For years, science ruled both my mind and my heart. As a child, I gravitated to science in school, mostly because it was one of the few things I could fully trust and rely on. Early on, I understood there were things that were definitive, certain, and undeniable in the natural world. Studying and learning about the nature of the physical and biological world provided me with a tremendous sense of fulfillment and security knowing there were things unfailingly true and real.

From the time I was five years old, I had clarity of purpose: *I knew I was going to be a doctor.* I wanted to be of service to others, and there was little that truly fascinated me the way the human body did and still does. The prestige of being a doctor also appealed to me. When a five-year-old says with confidence he's going to be a doctor, people stand up straight and look at him differently, with far more respect, and I reveled in that feeling as a child.

I dedicated myself to studying and pursuing that goal for years. Still, by the time I finished my undergraduate degree, there were deep psychological needs and questions within me that couldn't be answered and fulfilled through science alone. For many questions about life outside the classroom and the textbooks, I had no satisfying answers. And some of these are life's most important questions. Science was, and is, simply ill equipped to provide satisfactory answers to many of life's essential philosophical questions.

Science is, of course, an excellent resource for answering the questions of how and why many of the physical phenomena happening around us occur. But it's not so good at answering questions about how humankind should behave morally and ethically. Questions like: Why do we exist? Is there some purpose to life? Why does evil exist? Is it better to be noble and compassionate or to be more concerned with one's own needs? Does suffering have a purpose? What happens when we die? Why do we die? Do we live on after death? Does God exist? Does anything exist outside the universe? Such questions are just as important to understand as any scientific inquiry.

As I asked myself these and other essential questions, I began searching through various religions for answers. I returned to that empty feeling I'd had as a child as I tried but failed to understand why suffering is so prevalent and why apathy fills so much of humankind. I searched in philosophy to try to better understand the nature of humankind and the world, yet still found my mind and heart unfulfilled and incomplete as I tried to balance faith with reason.

As I studied the Bible, the Quran, and other religious and historical books and philosophies, I grew determined to know if there really is a God or some form of higher power that exists outside the realm of the scientifically explainable. Despite my search, nothing ever led me to any concrete evidence that gave me satisfactory answers. Religion is, simply, largely based on faith.

That period of philosophical and religious study proved invaluable once I began considering the nature of unseen entities. I read many accounts of spirits that I found intriguing from religious texts and books on the occult. Do beings that have no permanent physical form really exist? Is it possible that they have the enormous intelligence and power described and have interacted with humankind for thousands of years? These questions seemed to belong more appropriately to the realm of fantasy, science fiction, or religious philosophy. They didn't seem like something a scientifically oriented person should give any significant consideration to. For years, I dismissed these accounts as insignificant until my own transformative experience.

I had an epiphany and asked myself a fundamental question I'd never considered before my experience: Why does nearly every culture that has ever existed have some form of medicine man or shaman who claims to interact with spirits for the benefit and welfare of their community? As I considered this question, I began reading more about shamans and various cultures that were, and perhaps still are, separated by both vast geographic distances and centuries of time in origin. I found similarities in the methods, characteristics, attitudes, and practices of various shamans despite a significant degree of cultural and geographic difference between them. It's obvious such similarities are beyond mere coincidence. Either the accounts of shamans and the unique abilities ascribed to them are accurate, or nearly every people who ever lived suffers from the exact same delusion.

Though it is not always the case, in nearly every culture in which a form of shamanism exists, there are patterns of characteristics, practices, and experiences commonly displayed among those claiming to communicate with unseen entities for the healing of their community. Here are some of the most frequent similarities.

1. *The society.* Shamans live in a society or community that accepts that unseen entities are real and acknowledges that

the shaman is endowed with special abilities as a result of their ability to communicate and interact with these entities.

2. *The person.* The shaman frequently has an eccentric personality and, at times, though not necessarily, unique physical characteristics like an abnormal number of teeth or digits or some other atypical bone structure—but this is not universal. There is nearly always a hyper-developed sense of intuition and sensitivity demonstrated by the shaman that helps to distinguish them from others in the community.

3. *Mediumship.* Shamans have the ability to communicate with unseen entities, oftentimes directly but also through the ability to enter trance-like states where they are able to interact with the entities and/or channel the entities through their bodies. When going into these trance-like states, as well as while healing and performing other rituals, the shaman often uses the assistance of certain objects such as bull horns, feathers, a large sea turtle shell, drums, a glass of water, headgear, a gown, a metal rattler, a mirror, a staff, etc. The specific shapes and properties of these instruments are useful in various ways, depending on the specific form of shamanism practiced.

4. *The guide.* Shamans are often assisted by an active guardian entity or group of unseen entities that have distinct personalities and interests.

5. *Resistance to the call.* The unique and powerful abilities of the shaman are believed to result from a choice made by one or more unseen entities. The one who is chosen—often while still a child or adolescent—may resist this calling, sometimes for years. Harassment and/or torture by the entity, appearing in the form of physical or mental illness inflicted on the person or someone they love, breaks the resistance to the entity or entities' call, and the shaman accepts the calling to work with the entity or entities as a healer.

As mentioned previously, the impetus for writing *The World Hidden* initially came from my own experience with a shaman, a santera who lives in a small province near where my wife grew up in Cuba. For some, referring to a santera as a shaman is a significant inaccuracy. Still, I chose to use the term *shaman* in the book to refer to any healer who interacts with spirits for the sake of healing. I did this as a way to help the reader unacquainted with the world of spirit-guided healers. However, I am aware of the significant cultural and religious belief system differences between various individuals who interact with unseen entities.

My experience led me to seek to understand more about Santería, the religion that the santera practices. While Santería has its own rules and rituals, and its practice is predominantly rooted in the Caribbean, parts of the US, and South America, I immediately felt that there would likely be others from various other cultures around the world who work with spirits. I'd noticed this studying religious texts from various cultures. I viewed diverse global shamanic figures back then, and still do to this day, as various branches on a tree that share a common root, despite there being differences in the various beliefs and rituals individual shamans use.

I left Cuba convinced that unseen entities are real. Though certain unseen entities, often referred to as spirits, are real, I still had no clue how to explain these entities scientifically. Everything I had learned about unseen entities before my own experience came from religious and occult texts and movies. I had no other point of reference to unseen entities beyond the influence of my personal religious and cultural influences. But are religious and occult books really the definitive experts on spirits? To try to scientifically and objectively characterize the nature of unseen entities based on guidance from religious and occult texts, some of which within themselves claim to be "inspired by God," creates a number of challenges—not the least of which is, does God actually exist? That is far too lofty a question for me to answer, for it's certainly scientifically unprovable at present, so I decided to focus on something that seemed more accessible and reasonable. Could it be that what people commonly refer to as spirits, entities often viewed only through the lens of religious mysticism, fear, and faith, are very real intelligent life forms existing outside a physical form we can easily comprehend? Could it be that Indigenous and other peoples from times long past, including the writers of the Bible and other religious texts, were describing these entities in religious terms because they had no other lexicon or perspective with which to explain them?

As I began considering the nature of unseen entities, I reluctantly accepted that the primary sources of literature on the subject were religious books and books on the occult, neither of which would be met with a significant amount of credulity or reliability by scientifically oriented minds. Yet, regardless of that ambiguity, I decided to consider these texts anyway and see if perhaps there were nuggets of truth in them that would help me in my consideration of the possibility that unseen entities are real. After all, at least religion and the occult are aware that unseen entities exist, which shows that, in some respect, they are much further along than our current scientific understanding, which doesn't even acknowledge the reality of these entities at all.

For the vast majority of cultures in nearly every epoch of human existence, shamans have existed in various forms. In truth, however, there are significant differences in the belief systems of individuals who interact with spirits. For example, cultural heritage and religious framework affect the way a santera uses rituals to interact with unseen entities in comparison with, say, how a Buddhist shaman interacts and works with them. The individual spirits a santera may interact with are frequently specific Orishas, that is, beings with specific likes, interests, and personalities. The Buddhist shaman, however, may not work with a widely known or popular spirit but rather one familiar to him only.

Nevertheless, despite ritualistic and belief system differences, there are many similarities that can be found in the manners and practices these and other individuals who interact with unseen entities share. For example, while a shaman from a Native American tribe may use feathers or animal skulls or bones as tools in rituals and interactions with unseen entities, a santera may use a large sea turtle shell or dolls as a way to connect with them. While the specific objects may be different, there is still a common theme in that similar objects are used; at times, the same objects are used. Shamans also share other tools, like the use of chicken, pigeon, goat, or other animal blood in rituals.

Another example of similarity in rituals is found in the use of local herbs, such as peyote in North America and ayahuasca in South America. While very different pharmacologically, both of these substances fling their users into hallucinogenic states with considerably more lucidity than, say, PCP or ketamine. Were the shamans using these substances and providing them to their communities in rituals merely trying to get people stoned so that they could deceive them into believing spirits are real? That certainly is one potential point of view. But what do people who've actually participated in such rituals describe? Certainly not some kind of hallucinogen-driven party; rather, they describe deeply spiritually enriching experiences that provide some sort of timely direction and guidance needed in their lives.

Convinced as I was by the authenticity of the *santera* I met and the awareness that unseen entities do exist, a fire lit inside me to try to understand their nature. I knew that there would be other shamans and people who have consulted with them that likely had experiences similar to my own.

My wife grew up in Cuba, an island nation where Santería is practiced heavily. Santería is a religion where the existence of unseen entities not only is openly acknowledged but also plays a central role in the practice of the religion. Rituals are used to gain favor with specific entities, and priests of the faith claim to both see and communicate with those entities. Acts of healing, divination, and guidance are often provided to assist those in the presence of the priest interacting with an entity or entities.

I began talking with my in-laws and friends in Cuba about my experience. In return, they shared some of their own experiences with curanderos, santeras, and, in some cases, paleros. I came to see that there was a wealth of opportunities for me to receive answers to some of my many questions within the community of friends and family I have in Cuba. So, I sat down with them. Their accounts and points of view fascinated me. I became convinced that their accounts of unseen entities and shamans had tremendous value to the public, who are largely unfamiliar with these entities and what they can do. I then met other individuals and listened to their experiences. I met with reputable santeros, curanderas, shamans, paleros, and others who interact with unseen entities around the whole of Cuba and the United States, particularly the American Southwest. We, too, sat down and talked. I grew even more convinced that their experiences had merit and should be considered by the scientific community and the public in general.

I live and work in New Mexico near several Native American reservations. There are many practicing shamans, curanderos, and others who interact with unseen entities in the area. I've been fortunate enough to associate with some of these individuals and enjoy their cooperation in sharing their experiences with unseen entities. The accounts detailed in *The World Hidden* describe several common abilities that cultures with shamans around the Earth describe as manifesting as a result of these individuals' relationships with unseen entities.

CLAIRVOYANCE

Clairvoyance is the ability to perceive events in the future or things beyond normal sensory contact and to see and/or hear unseen entities around them. Nearly every known culture that has or had shamans reports that shamans have this ability. For some shamans, clairvoyance comes from direct communication with unseen entities. For others, it occurs through the recognition of signs, physical phenomena occurring around them, or some organized form of divination, such as *caracoles*, which indicates what a future outcome may or will be. Clairvoyance occurs through varied and diverse means, whether through direct communication from unseen entities or what can be accurately described as a hyperdeveloped sense of intuition. There is often great variability in the way different individuals who interact with unseen entities perceive clairvoyance. Yet in each individual's case, that experience is often associated with a definitive clarity that allows that particular individual to understand the degree to which some event they are physically removed from is happening or will happen and/or its degree of likelihood of occurring in the future.

HEALING

Healing is the ability to gain knowledge of physical and psychological ailments through the guidance of unseen entities. Unseen entities often direct the solution to the ailment through various means. At times, herbal preparations are used. At other times, prayer, blowing smoke, and energy cleansing with leaves and/or an egg are all that are needed. In cases of more serious health problems, animal sacrifices, including the use of blood, are required to accomplish healing.

MEDIUMSHIP

While not all shamans and others who interact with unseen entities channel them into their bodies, many do. This ability to channel these entities often results in the manifestation of various abilities from these entities, or a more potent expression of them. For example, acts of healing, clairvoyance, and/or a clearer perception of the problem(s) affecting the individual seeking assistance occur more intensely when an unseen entity is channeled through the shaman. Terms such as *in a trance* and *in transition* are often used to describe periods when a shaman or other person is actively channeling an unseen entity through them.

For most mediums, there is a gradual process that occurs over time that helps them to understand what they can do with these entities. Frequently, a person begins to see unseen entities in childhood (i.e., an imaginary friend), and the entity or entities engage in activities to communicate with the person, often through direct conversation and repeatedly visiting or touching the person, often while they are resting or attempting to sleep. Often, there is an initial reluctance on the part of the person with the ability to become a medium and accept that unseen entities are real. There is frequently a period of refusal on the part of the individual, during which the entity forces the person to accept what they are and to work with the ability they have been given to be of service to others. This coercion is often achieved by the entities making either the shaman or someone close to them ill and then providing the healing to the one affected when the shaman accepts their gift and follows the entities' direction. Many of the accounts in *The World Hidden* illustrate this pattern.

THE ABILITY TO AFFECT OUTCOMES

A fascinating ability both shamans and others manifest through their interactions with unseen entities is the ability to direct favorable and/or harmful outcomes to others. Shamans generally work to accomplish positive outcomes. At times, the effects of protection rituals can harm someone seeking to harm an individual or individuals the shaman has performed a protection ritual with. In so protecting an individual or individuals through their work with unseen entities, the harmful outcome may return to the individual seeking to harm someone protected. The shaman often views this consequence as the natural outcome of the malevolent intent of the person seeking to do evil.

This ability to affect outcomes commonly pertains to health, but also, through their interactions with unseen entities, shamans and others can affect legal outcomes or encourage individuals to take actions they might otherwise resist or reject in matters of love, business, politics, and many other arenas. Frequently, a person working with unseen entities to bring harm can similarly accomplish the desired outcomes through direction from unseen entities to affect the person targeted; through violence the entities can encourage in others; or through accidents, theft, and other actions. The manner and mechanisms of such unique activity are fascinating and engender a tremendous number of legitimate questions regarding how unseen entities can affect these actions. Many accounts reference these realities.

After considering such accounts and personally experiencing similar things in my dealings with shamans, I was moved by the depth of humility that I've found in these individuals. Even the noblest person can be easily tempted to develop feelings of superiority and arrogance when endowed with common human advantages like wealth, intelligence, and prestige. Yet the overwhelming sense of humility and modesty in shamans, who are often endowed with the ability to achieve incredible feats of healing, clairvoyance, and other deeds, begs the question: Why such modesty?

Such modesty has two obvious and immediate causes, which are readily discernible. First, the individuals gifted to interact with unseen entities are often those with an appetite for what are often considered spiritual things. The shamans, curanderos, santeros, and other practitioners I know who use their work with unseen entities for benevolent actions pride themselves on having character of the highest caliber. They often view such character as inherent to their ability to maintain favor with the unseen entities that endow them with the extraordinary gifts they manifest as well as an awareness that their abilities are exclusively the result of sources outside themselves. They understand that such works have nothing at all to do with any unique ability within themselves but are the fruitage of their partnerships and relationships with the unseen entities whose strength and power flow through them.

Second, like those who use their relationships with unseen entities for less selfless reasons, the modest ones understand, often from personal experience or through the direct observation of others, that when the unseen entities are not honored or given the respect and recognition they merit, the flow of power from them is stopped. These entities can also be angered into harming a person who once was allowed to channel the entity's ability if that person fails to honor some promise or conduct themselves in ways the entity feels are appropriate.

I realize that for many of the individuals who work with unseen entities, the use of the term *alien life forms* as it relates to what are commonly known as spirits is not ideal. Opinions vary among shamans and others concerning the origin and nature of the entities they interact with. For some, spirits represent nothing less than deities and are, in fact, prayed to and worshiped, whereas, for others, spirits are less than gods and more like benevolent helpers. Still others see spirits as a form of life that lives on after our physical body has died. Beyond these few opinions, there are still many other beliefs. The spirits of the ancestors, for example, are a common belief among many African peoples and various Indigenous cultures around the world. It is not due to a lack of respect that I refer to unseen entities as alien life forms; it is with an awareness that unseen entities are, in fact, intelligent nonhuman entities whose origin and exact nature remain disputed even among shamans. This lack of congruence in thought among those who interact with unseen entities is another reason why further study is necessary.

So, I invite you into the realm of some of the world's true alien life forms. Not only are unseen entities real, but they are enormously intelligent beings in possession of powerful abilities and understanding far beyond humankind. For centuries, their presence has been hidden in the realm of the mysterious and religious, hiding in plain sight a truth so obvious and yet one only a few, relatively select, gifted individuals among humankind have come to know intimately. Alien beings have watched, communicated with, and interacted with humankind for thousands of years.

An Epilogue of Sorts

On Spirits was the culmination of five very difficult years of study, interviews, observations, and explorations. I am very grateful to my wife for her extreme patience. My near obsession with trying to make a contribution of significance to humanity can be exhausting and irritating. Through it all, I've tried to maintain a sense of religious neutrality the best I could. Such was not particularly easy given the subject matter. I was examining the existence of unseen entities, and the leading experts on the subject are religious scholars and writings. I discussed and reviewed the written works of various Eastern and Western religious persons, Indigenous ethnographers, and esoteric practitioners. Herein lies what I hope might be one of the more important consequences of what I expect will be my most important nonfiction scientific literary work. *We have to turn down the rhetoric and just talk to one another.* We live in a world of extremes where opinions vary broadly, particularly with regard to political and religious thought, and such divergent opinions can at times even engender violent outcomes. My success, if my work might even engender such a thought, was largely the result of the combined effort of study, research, and good old-fashioned journalism. It would not have been possible were it not for the unselfish assistance of those with vastly different religious practices and thoughts than my own. I thank them again for their forthright candor and help.

As a man strongly devoted to my own Christian beliefs, and a person appreciative of scientific theory, I've attempted, hopefully successfully, to maintain objectivity and avoid the flippant characterization of the belief systems, cosmology, and ethics of individuals and cultures I, at least initially, lacked a significantly nuanced understanding of. I sincerely try to live conscientiously in harmony with the guidance found in God's word, the Bible; however, I humbly acknowledge I've not always done so, nor been entirely consistent in applying biblical counsel. And while I have opinions about the morality of many of the practices I've outlined within the body of this work, I find Jesus' strict words not to judge others both scientifically and spiritually beneficial.

I forthrightly admit, I do believe firmly in a singular divine person, a Supreme Being, a Great Spirit, Elohim, Yahweh, Allah, Jehovah. So strong is my belief, I feel with 100-percent certainty, this person both exists and is the rewarder of those earnestly seeking him. And despite his clear feelings about spiritism as outlined in the Bible, I recognize that there are individuals and cultures for whom such practices, though clearly offensive and contradictory to what the Bible describes as behavior approved by God, are done routinely with a clear conscience. I am aware that there are many with noble intentions who engage in such conduct. I am also aware that such often leads to outcomes that affect people's lives in ways many would feel are positive. In reporting such accounts as a journalist, and seeking to understand the scientific forces associated with these, I know I am not engaging in conduct that is either unethical or spiritually bereft. The Creator, or Great Spirit as some call him, is one who cherishes both truth and free will. He has given humankind an opportunity to understand his creation and the freedom to pursue whatever conduct one may choose to pursue. Studying such practices, and the cultures and individuals engaging in such, proves the existence of unseen intelligent beings and is itself a declaration of the veracity of various truths found in the Bible and other religious texts. These, as well as accounts from Indigenous people, have accurately, at least to some degree, described the existence of unseen intelligent beings long before Western science identified them. The cosmology, spiritual practices, and beliefs of major religious groups, Indigenous peoples, and other esoteric belief systems offer humanity a window into the collective wisdom of generations of humans long past. Science has much to learn from these ancient peoples and their accounts, and sorting legitimate observations of natural phenomena from so-called superstition is, in my opinion, as worthy a scientific endeavor as any other.

Children often make the mistake of thinking they are wiser than their parents. Humanity often imitates such thinking, believing the individuals within modern society are somehow wiser and more intelligent than our ancestors who came before us. This is a gross error. Though there has been tremendous scientific and technological progress, such has not contributed to a more equitable or sustainably stable and prosperous human society, nor has man's technological advancement resulted in a superior understanding of the unseen spiritual realities and beings that affect our world. And as such, humanity has much to learn from our ancestors in this regard.

I cannot say what path *On Spirits* will move humanity toward, if any. I no longer need to understand such or even care to. Humility beckons I give no significant attention to such ego-engendering cogitation. I've done the best I can to elucidate what's been a great mystery for most of modern humanity, and I feel a measure of gratitude for having done my part to the best of my ability with the help of my Creator. "Your God is the revealer of secrets," said Nebuchadnezzar to the prophet Daniel when that man of God revealed his unique dream to him. Such is my own feeling regarding my work, and I know for a certainty, my discovery was not accomplished without the benevolent help of my loving Heavenly Father and God.

I do sincerely hope *On Spirits* engenders a measure of faith in the divine. It seems obvious such beings' existence supports the existence of the divine, though I candidly cannot prove such. While no human can speak with complete accuracy about the scientific nature of unseen entities, the Bible and other religious texts offer an explanation that is at least worth considering, especially given that these texts identified such beings millennia ago. It is my sincere hope science will continue to move toward a more objective perspective regarding religious and ethnographic texts as such pertain to cosmology and unseen entities. I firmly believe, prudently assuming at least some accuracy on the part of such texts regarding such matters, humanity will move further along in its scientific understanding of them. Verification of the existence of unseen entities affirms the wisdom of abandoning the long-held ignorance that such texts and accounts have no scientific merit. May humanity be forever changed in accepting what our ancestors have known for millennia: we are indeed not alone in the universe.

Thank you for reading *On Spirits*. If you'd like to see the photography and other works by Dr. Joseph Michael Feagan,

PLEASE VISIT
www.TheWorldHidden.com

Glossary

Altar—Shamans, curanderos, santeros, paleros, espiritistas, and others who routinely interact with spirits usually have an altar. The altar serves as a focal point where the person is able to interact more directly with the spirit or spirits they have a relationship with. The altar usually contains some physical representation of the spirit or spirits the individual works with and also a means by which they can make offerings to the spirits. Depending on the spirit's desires, it is common for things like toys, metal, cigars, rum, sweets, and other things to be given in exchange for the assistance and support of the spirit or spirits interacting with the person.

Babalao—An initiated priest of Santería. Babalaos are particularly gifted in divination and use a variety of methods for practicing divination.

Bembé—A ritual of Santería that includes prayer, singing, praising, drums, and other musical instruments. Often divination, wisdom, and acts of healing are performed during the course of the ritual led by a priest or devotee of a particular Orisha.

Brujería—Witchcraft.

Consultee—A person who goes to see a shaman or medium or someone else who interacts with spirits.

Curandero—A person who works with spirits for healing.

Espiritista—A person who works with spirits, often serving as a medium channeling spirits through their body.

In transition/In a trance—Individuals who work with spirits often experience a period during which they allow spirits to enter their own bodies for the assistance of others. At times, they also assist others by guiding spirits into or removing spirits from individuals. The period when the spirit enters a person is described in various terms. *In transition* and *in a trance* are two of the more common terms used to refer to the period when a spirit or spirits are inside a person's body. Often during this period, the person in transition cannot recall the words spoken or deeds performed while the spirit is inside them.

La Regla de Ocha—The Spanish term for Santería.

Santero/a—A priest or priestess of the Santería faith. Santeros may have a variety of gifts through their interactions with spirits. Divination, clairvoyance, healing, and channeling spirits as mediums are not uncommon for santeros. Some santeros may channel spirits in each of these ways, while others may only have the gift of healing. Some santeros may also be particularly gifted at a specific type of healing, for example, helping pregnant women or children.

Limpia—A cleansing ritual done to remove the bad energy from someone.

Orishas—The spirit deities of the Yoruba religion. The Orishas are revered in Santería, Voodoo, and other religious practices derived from the religion of the Yoruba people.

Palero—A practitioner of the "el Palo" belief system. The Palo belief system is based on two main pillars: the veneration of spirits and the belief in natural terrestrial powers. All natural objects, and particularly sticks, are believed to be infused with powers, often linked to the powers of the spirits.

Santería—A religion practiced predominantly in Cuba, South America, and the USA. Santería developed from the Yoruba religion of slaves from West Africa and combined many beliefs of Catholicism and Indigenous Cubans. The religion centers around the worship of the Orishas and other deities synchronized with Catholic saints.

The Bad Eye—Many cultures believe in the power of the "evil eye," a kind of curse that can bring injury or harm to a person or household receiving envious, jealous, and malevolent wishes from others. Children are thought to be particularly susceptible to the effects of the bad eye. Children are often given necklaces or bracelets to protect from *the bad eye*. Adults also use necklaces, pendants, amulets, or other objects for a similar reason.

The Virgen—La Virgen de la Caridad, the Virgin Mary.